AFRICAN SWINE FEVER

DEVELOPMENTS IN VETERINARY VIROLOGY

Yechiel Becker, Series Editor
Julia Hadar, Managing Editor

Payne, L.N. (ed.) *Marek's Disease* (1985)
Burny, A. and Mammerickx, M. (eds.) *Enzootic Bovine Leukosis and Bovine Leukemia Virus* (1987)
Becker, Y. (ed.) *African Swine Fever* (1987)

AFRICAN SWINE FEVER

edited by

Yechiel Becker
The Hebrew University of Jerusalem
Israel

Martinus Nijhoff Publishing
a member of the Kluwer Academic Publishers Group
Boston/Dordrecht/Lancaster

Distributors for North America:
Kluwer Academic Publishers
101 Philip Drive
Assinippi Park
Norwell, MA 02061 USA

Distributors for the UK and Ireland:
Kluwer Academic Publishers
MTP Press Limited
Falcon House, Queen Square
Lancaster LA1 1RN UNITED KINGDOM

Distributors for all other countries:
Kluwer Academic Publishers Group
Distribution Centre
Post Office Box 322
3300 AH Dordrecht
THE NETHERLANDS

The figure on the cover is from Vinuela, E., "Molecular Biology of African Swine Fever Virus." The figure appears on page 37 of this book.

Library of Congress Cataloging-in-Publication Data

Becker, Yechiel.
    African swine fever.

    (Developments in veterinary virology)
    Bibliography: p.
    Includes index.
    1. African swine fever. 2. African swine fever virus.
I. Title. II. Series.
SF977.A4B43   1987   636.4'0896979   86-23851
ISBN-13: 978-1-4612-9437-5   e-ISBN-13: 978-1-4613-2343-3
DOI: 10.1007/978-1-4613-2343-3

CONTENTS

CONTRIBUTORS

M. Ackermann
Federal Vaccine Institute
4025 Basle
Switzerland

Yechiel Becker
Department of Molecular Biology
Faculty of Medicine
The Hebrew University of Jerusalem
P.O. Box 1172
91010 Jerusalem
Israel

William R. Hess
United States Department of Agriculture
Agricultural Research Service
Plum Island Animal Disease Center
P.O. Box 848
Greenport, New York 11944
USA

Ueli Kihm
Federal Vaccine Institute
4025 Basle
Switzerland

Miguel Angel Marcotegui
Department of Animal Virology
Instituto Nacional de Investigaciones Agrarias
Embajadores, 68
Madrid 28012
Spain

Charles A. Mebus
United States Department of Agriculture
Agricultural Research Service
P.O.Box 848
Greenport, New York 11944
USA

H. Mueller
Federal Vaccine Institute
4025 Basle
Switzerland

Angel M. Ordas Alvarez
Department of Animal Virology
Instituto Nacional de Investigaciones Agrarias
Embajadores, 68
Madrid 28012
Spain

VIII

I.C. Pan
Veterinary Medical Officer
U.S. Department of Agriculture
ARS, NAA
Plum Island Animal Disease Center
P.O. Box 848
Greenport NY 11944
USA

R. Pool
Federal Vaccine Institute
4025 Basle
Switzerland

Jose Manuel Sanchez-Vizcaino
Instituto Nacional de Investigaciones Agrarias
Department of Animal Virology
Embajadores, 68
Madrid 28012
Spain

Enrique Tabares
Departamento de Microbiologia
Facultad de Medicina
Universidad Autonoma de Madrid
Arzobispo Morcillo 4
Madrid 28029
Spain

Catharinus Terpstra
Department of Virology
Central Veterinary Institute
P.O. Box 365
8200 AJ Lelystad
The Netherlands

Eladio Vinuela
Centro de Biologia Molecular (CSIM-UAM)
Facultad de Ciencias
Universidad Autonoma
Canot Blanco
28049 Madrid
Spain

PREFACE

African swine fever (ASF) is caused by a virus that is classified as a member of the Iridovirinae family. The disease in the warthog, the natural host, in Africa was described in 1921 by R.E. Montgomery. The reservoir of the virus is in ticks. The introduction of domestic pigs into territory occupied by warthogs infected with ASF in the 1960's has endangered the pig industry around the world. The domestic pig is highly sensitive to ASF and develops a devastating disease that kills the pig without giving the immune system a chance to defend the animal against the virus infection. The ability of ASF virus to infect and destroy cells of the reticuloendothelial system leaves a defenseless host that succumbs to an infection which may be described as an acquired immune deficiency disease of domestic pigs.

Introduction of the virus into Iberia in the 1960's led to a series of ASF epidemics in Spain and Portugal, and later in France, that caused heavy economic losses. Between 1978 and 1980, ASF virus made its appearance in Malta and Sardinia, as well as in Brazil, The Dominican Republic, Haiti, and later in Cuba. In 1985-6, ASF appeared in Belgium and The Netherlands. In the absence of an effective vaccine, the only way to stop the spread of the virus among herds is to slaughter all pigs in the infected regions. This is a costly operation which has marked economic effects on meat production and on commerce.

The present volume in the series DEVELOPMENTS IN VETERINARY VIROLOGY deals with African swine fever from various points of view. The clinical and pathobiological aspects of the disease, as well as pathogenesis of the infection, are presented. The molecular biology of the virus has been detailed, since analysis of the viral DNA and proteins will eventually provide an understanding of the virus genes involved in the disease. The differential diagnosis of ASFV is presented, as are the current approaches to the development of an effective vaccine to protect pigs against ASF. The epidemiology of the virus, also described in this volume, shows that ASFV is indeed a major threat to the pig industry around the world.

This volume is dedicated to all the scientists, veterinarians and the veterinary services whose task it is to eradicate this disease and who maintain a constant watch to prevent its spread in their respective

X

countries. It is hoped that the material contained herein will be of help to all of them in their endeavors. I wish to thank all the authors for their cooperation and contributions.

This book is a tribute to Mr. Raymond Craps, Director VI/F-4, Directorate-General for Agriculture, Commission of the European Communities, Brussels, Belgium, in recognition of his two decades of devotion to research on virus diseases of concern to agriculture. The continuous support by the Commission of the European Communities made possible many of the research developments on African Swine Fever described in this book.

Yechiel Becker

Jerusalem

AFRICAN SWINE FEVER

1

INTRODUCTION

W.R. HESS

African swine fever (ASF) is a complex and devastating disease of
domestic swine. It has occurred in forms ranging from peracute to
acute, subacute, chronic and inapparent and with mortality rates
ranging from close to 100% to as little as 3%. Its causitive agent is
a large icosahedral cytoplasmic DNA virus, and as such has been
tentatively placed in the family Iridoviridae (1). It is the only
member of the family that infects a mammal, and it is the only DNA
virus presently known to satisfy the criteria for classification as an
arbovirus (2). Although it has undoubtedly been present in soft ticks
and the indigenous wild swine in Africa for a very long time, it did
not emerge as a disease agent until breeds of domestic swine from
Europe were brought into Africa. Then, that which existed unbeknown
as an inapparent infection in warthogs now appeared as an acute,
highly contagious and lethal febrile disease of domesticated swine
that was readily mistaken for acute hog cholera (HC). Montgomery
described the disease in a report published in 1921 that recounted his
work on ASF in Kenya during 1909 through 1915 when fifteen outbreaks
occurred involving 1366 pigs of which 1352 or 98.9% died (3). In
addition to describing the disease, he established its viral nature,
studied the transmission of the virus and its survival under a variety
of environmental conditions, explored methods of immunization, studied
the host range and indicated the possible role of wild swine in
maintaining the virus in nature. However, he believed the virus to be
another serological type of the causitive agent of HC. Despite the
tremendous advances in technology that have since occurred, some basic
problems that Montgomery was unable to resolve still remain. The
immunology of the disease continues to be obscure, and a satisfactory

vaccine has not been developed.

Outbreaks of ASF were soon encountered in a number of other areas south of the Sahara in Africa, and the studies that ensued confirmed and enlarged upon Montgomery's findings. Warthogs (Phacochoerus aethiopicus) and bush pigs (Potamochoerus porcus) were found to be inapparent carriers of the virus (4, 5), but the means of transmission of the virus from the wild to the domestic pig was not readily apparent. Arthropods ranging from fleas, lice and ticks to blood-sucking flies were suggested and in some instances tested as possible vectors, but the settlers in East Africa found that swine could be quite safely and profitably produced by confining them to pens that isolated them from the wild pigs. While the number of outbreaks of ASF in East African was greatly reduced, elsewhere in Africa where such methods were not adopted, the disease became established in domestic pig populations, and considerable losses resulted. In the Western Cape Province of South Africa in 1933, the disease spread among 11,000 domestic pigs without intervening passages in wild pigs (6). Eight percent of the pigs survived, and as suggested by Scott (7), this was perhaps the first indication that the virus and its new host were beginning to adapt to each other. However, continued adaptation was halted in South Africa by a slaughter program that quickly eliminated all of the surviving pigs in the affected area. In Angola where domestic pigs were still allowed to range freely, the swine industry was severely impaired by the disease. Subacute and chronic ASF began to appear with greater frequency. This was thought to be the result of modification of the virus through continued association with domestic pigs (8).

Aside from a greatly enlarged and engorged spleen frequently seen in pigs that had died with the acute disease (3, 5, 6,), there were no lesions that distinguished ASF from HC. Challenge inoculation of pigs immunized against HC was then the only means of distinguishing the two diseases. Occasionally domestic pigs were encountered that had survived the infection and were refractory to challenge with the homologous virus isolate, but the numbers were never large enough to enable extensive studies to be undertaken (4, 5, 6,). Because the animals seldom withstood challenge with an ASF virus isolate from a different outbreak, it was believed that many immunologic types existed. This was difficult to prove. Virus neutralizing antibodies were never demonstrated, and surviving pigs were too few to enable extensive cross immunity trials to be conducted. Also, the survivors were invariably found to be virus carriers. All efforts to produce effective vaccines with virus rendered noninfectious by various chemical or physical means failed.

Because the incidence of subacute and chronic infections was relatively low, attention centered largely on the more devastating forms of the disease; therefore, ASF was usually described as a peracute, fulminating disease with mortality close to 100 %. Nevertheless, most of the major swine producing countries were not greatly concerned until the disease appeared in Portugal in 1957 and was eradicated but reappeared in 1960 and spread to Spain. At the outset, the disease on the Iberian Peninsula fit the description of ASF as it was usually encountered in Africa. In the first incursion of ASF in Portugal, 17,000 pigs were lost before the disease was eradicated. At about the same time, modern research on ASF was ushered in by the discovery of the hemadsorption reaction and the adaptation of ASFV isolates to growth in cell cultures (9, 10). It was now possible to quite rapidly and economically diagnose the disease and differentiate it from HC, and the virus could be accurately assayed and produced in large quantities without the costly use of pigs. Although an increased incidence of subacute and chronic ASF was noted in Spain in the early sixties, the devastating acute disease still predominated, and was the form involved in the early incursions of ASF in France and Italy.

Malmquist and others (10, 11, 12, 13) found that by passing ASFV isolates serially in cell cultures, virus strains could be derived that were no longer capable of producing clinical ASF, and pigs inoculated with such strains could withstand challenge with the fully virulent homologous parent virus isolate. In Portugal, virus that had been serially passed in swine bone marrow cultures was used as a vaccine. Out of 550,000 animals that received the vaccine, 128,684 developed postvaccinal reactions including pneumonia, skin ulcers, abortions, locomotor disturbances, disturbances in lactation and death (11). Undoubtedly a large number of inapparent carriers were produced, and it has been speculated that some of the virus isolates of lower virulence involved in subsequent outbreaks may have emerged from this immunization effort (14).

On the Iberian Peninsula where ASF has been enzootic since 1960, the number of outbreaks has varied in a cyclic fashion over the years. One of the years of high incidence was 1977. It is perhaps more than coincidental that the disease appeared in several widely separated parts of the world the following year. Regardless of their origins, the vast majority of the virus isolates from those outbreaks have been ones of low virulence (15, 16), and many of the problems in controlling the spread of the disease have been complicated by this trend in the evolution of the disease.

REFERENCES
1. Fenner, F. Intervirology 7:91–100, 1976.
2. Plowright, W., Perry, C.T., Peirce, M. and Parker, J. Arch. ges. Virusforsch. 31:33–50, 1970.
3. Montgomery, R. J. Comp. Pathol. 34:159–191, 243–262, 1921.
4. Steyn, D.G. Onderstepoort J. Vet. Sci. Anim. Ind. 1:99–109, 1932.
5. DeTray, D.E. Adv. Vet. Sci. 8:299–333, 1963.
6. DeKock, G., Robinson, E.M. and Keppel, J.J.G. Onderstepoort J. Vet. Sci. Anim. Ind. 14:31–93, 1940.
7. Scott, G.R. Bull. Off. int. Epiz. 63:645–677, 1965.
8. Leite Velho, E. Bull. Off. int. Epiz. 46:335–340, 1956.
9. Malmquist, W.A. and Hay, D. Amer. J. Vet. Res. 21:104–108, 1960.
10. Malmquist, W.A. Amer. J. Vet. Res. 23:241–247, 1962.
11. Manso Ribeiro, J., Nunes Petisca, J.L., Lopes Frazao, F. and Sobral, M. Bull. Off. int. Epiz. 60:921–937, 1963.
12. Sanchez Botija, C. Bull. Off. int. Epiz. 60:901–919, 1963.
13. Hess, W.R., Cox, B.F., Heuschele, W.P. and Stone, S.S. Amer. J. Vet. Res. 26:141–146, 1965.
14. Hess, W.R. ASFV. Virology Monographs, Wien–New York. 9:1–33, 1971.
15. Wilkinson, P.J., Wardley, R.C., Williams, S.M. J. Comp. Path. 91:277–284, 1981.
16. Mebus, C.A. and Dardiri, A.H. Proc. U.S. Anim. Hlth. Assoc. 83:227–239, 1979.

AFRICAN SWINE FEVER VIRUS IN NATURE
W.R. HESS

Thus far, natural infections with ASFV have been found only in porcine species and ticks of the genus <u>Ornithodoros</u>. In Africa, the virus has been recovered many times from warthogs (<u>Phacochoerus aethiopicus</u>) and bush pigs (<u>Potamochoerus porcus</u>), and they have also been infected experimentally (1, 2, 3). A single isolation was made from a giant forest hog (<u>Hylochoerus meinerizhageni</u>) (4), and isolations from a hippopotamus (3) and a porcupine and hyena have also been claimed (5), but these findings have not been confirmed by additional isolations, nor have these species been infected experimentally.

Outside of Africa, other wild Suidae species have become infected or have been found to be susceptible to infection with ASFV. The European wild boar (<u>S. scrofa ferus</u>) has acquired the disease from domestic pigs, and its response to the infection has been clinically and pathologically similar to that of the domestic pig (6, 7). In Spain, a survey has indicated that contacts with wild boars account for 5.8% of the outbreaks in domestic pigs (8). In America, feral pigs of the southeastern United States have been shown experimentally to be highly susceptible to ASF with responses similar to those of the domestic pig (9). The American collared peccary that ranges from southwestern United States to Patagonia belongs to the Tayassuidae family and is susceptible to a number of virus diseases that affect domestic pigs, but it is resistant to ASF (10).

Animals other than porcine species that have been tested include cattle, horses, sheep, goats, dogs, cats, guinea pigs, rabbits, hedgehogs, hamsters, rats, mice and various fowl. With the exception of kids up to 4 months old (11), all were refractory to ASFV. The virus has been propagated in rabbits after a series of alternating passages

Y. Becker (ed.), *African Swine Fever.* Copyright © 1987. Martinus Nijhoff Publishing, Boston, All rights reserved.

in rabbits and pigs (12, 13, 11). It appears that porcine species are
the only vertebrates that may be naturally infected.

It was early recognized in Africa that ASF outbreaks usually
occurred in areas where infected warthogs existed, but the mode of
transmission of the virus from the warthog to the domestic pig was not
readily apparent. Since pigs placed in contact with infected warthogs
failed to become infected (3, 14, 15, 16), the possible involvement of
bloodsucking arthropods as vectors was explored. Montgomery found that
fleas and lice did not transmit the virus (1). Others confirmed that
the hog louse (_Haematopinus suis_) failed to transmit the virus by bite
(16, 17) but one study indicated that it was perhaps a mechanical
vector (18). Several hundreds of lice (_H. phacochoeri_ and _Haematomyzus
hopkinsi_) collected from infected warthogs in Kenya were tested, but no
virus was recovered (19). The suggestion that a flying insect might be
responsible for transmitting the virus from warthogs to domestic pigs
(20) seemed unlikely in view of the fact that "paddocking" of pigs
proved to be an effective means of protecting them from acquiring the
disease (3, 21).

Early attempts to incriminate ticks as vectors of ASFV in Africa
were unsuccessful (3). Ixodid ticks found on warthogs or in their
burrows failed to yield ASF virus, and _Rhipicephalis simus_ and
Amblyomma variegatum which represent 2 genera commonly found on
warthogs were unable to experimentally transmit the virus when fed
sequentially on infected and healthy susceptible pigs (19). Similar
results were obtained by others in attempting to demonstrate ASF virus
transmission with species of hard ticks (17, 22). Following Botija's
announcement that virus was recovered from _Ornithodoros erraticus_ found
on farms in Spain where ASF outbreaks had occurred (23), renewed
efforts were made to isolate virus from the _Ornithodoros_ species
commonly found in the burrows occupied by warthogs in East and South
Africa. These ticks were previously designated as _Ornithodoros
moubata_, Murray. Although the proper designations now are _Ornithodoros
porcinus porcinus_ (24) or _Ornithodoros moubata porcinus_ complex sensu
Walton (25), _O. moubata_ is used here for convenience to refer to the
eyeless tampan from warthog burrows which has been the main object of
ASF vector studies in Africa.

Experimental transmission of ASFV from infected to healthy
susceptible pigs by O. moubata was demonstrated (17), and in a series
of reports by Plowright and his associates, the potential of O. moubata
as a vector of ASF virus was revealed. Virus was isolated from ticks
collected from warthog burrows (26). Transtadial, sexual and
transovarial transmission of the virus in the tick were established
(27, 28). It has all the characteristics of a true biological vector,
and serves as a reservoir of the virus in nature. The close
association of the virus and the tick has prompted the suggestion that
ASFV may actually be of tick origin (19).

Although the ASFV-O. moubata-warthog complex is probably the most
important natural reservoir of ASF in Africa, it does not account for
many of the outbreaks on the African continent. In addition to
incursions of ASF that result from human activities such as the
movement of infected animals and the feeding of ASFV-contaminated
swill, there are outbreaks that appear to originate from unknown or
undefined reservoirs in nature. In an area of central Kenya at
altitudes of more than 6200 feet above sea level, no ticks were found
in any of the more than 100 warthog burrows examined, yet ASFV
infection in the warthogs was universal, and outbreaks in domestic pigs
occurred (19). On the other hand, there are parts of South Africa
where warthog populations appear to be entirely free of ASFV infection
(21). In other areas where O. moubata and the warthog coexist and both
populations are carrying the virus, there may be substantial
differences from one geographical location to another in the overall
infection rates in the two populations, the infection rates in various
age groups of the warthogs, the maximum virus titers found in each
group and the percentage of warthog burrows yielding infected ticks
(19). There has been much study and speculation concerning the reasons
for these differences (19, 30). However, if we consider the number of
ways that ASFV isolates may differ from one another and the differences
that may exist in a single tick species collected from different
geographical locations, variations in the ASFV-O. moubata-warthog
complex and its epizootiological manifestations are to be expected.

Since ASFV has not been detected in warthog excretions, nor has
evidence of transplacental or milk transmission been found, it is

probable that another arthropod vector is involved in the spread of the virus in areas where warthogs are infected in the absence of O. moubata. The establishment of O. erraticus as a reservoir for ASFV in Spain clearly indicates that O. moubata is not an unique arthropod vector for the virus. In fact, recent studies indicate that most, if not all, Onithodoros species that will feed on mammals may be capable of acting as vectors of ASFV. Thus far, 3 species in North America and the Caribbean Basin, O. coriaceus, O. turicata and O. puertoricensis have been shown experimentally to be susceptible to infection with ASFV and to be capable of transmitting the virus by bite to healthy susceptible pigs (31). O. savignyi, a widely distributed species in Africa , has likewise been found to be experimentally capable of acting as a biological vector of the virus (32). Since it is found in many areas where ASF occurs, it may actually be serving as a natural vector of the disease. None of the ixodid ticks that have been tested have been found capable of transmitting ASFV; however, only a few species have been tried. Others that are known to be vectors of several diseases should be tested before the ixodid ticks are completely ruled out as potential vectors. In addition to ticks, fleas and lice, there are a number of blood-sucking insects that are yet to be tested. A number of new approaches to the control and eradication of insect populations are being developed. Most require a thorough knowledge of the biology of the particular arthropod species concerned. Such information could be of tremendous importance in designing means of controlling and eradicating ASF.

REFERENCES
1. Montgomery, R. J. Comp. Pathol. 34:159-191, 243-262, 1921.
2. Steyn, D.G. Onderstepoort J. Vet. Sci. Anim. Ind. 1:99-109, 1932.
3. DeTray, D.E. Adv. Vet. Sci. 8:299-333, 1963.
4. Heuschele, W.P. and Coggins, L. Bull. Epiz. Dis. Afr. 13:255-256, 1965.
5. Cox, B.F. Bull. Epiz. Dis. Afr., 11:147-148, 1963.
6. Polo Jover, F. and Sanchez Botija, C. Bull. Off. int. Epiz. 55:107-147, 1961.
7. Ravaioli, L., Palliola, E. and Ioppoto, A. Vet. Ital. 18:499-513, 1967.
8. Ordas. A., Sanchez Botija, C. and Diaz, S. in: Proceeding of a CE/FAO research seminar, Sassari, Sardinia, 23-25 Sept. 1981, pp. 7-11, EUR 8466 EN, 1983.
9. McVicar, J.W., Mebus, C.A., Becker, H.N., Belden, R.C. and Gibbs, E.P.J. J.A.V.M.A. 179:441-446, 1981.

10. Dardiri, A.H., Yedloutschnig, R.J. and Taylor, W.D. Proc. U.S. Anim. Hlth. Assoc. 73:437-452, 1969.
11. Kovalenko, J.R., Sidorov, M.A. and Burba, L.G. Bull. Off. int. Epiz. 63:bis:169-189,1965.
12. Mendes, A.M. and Daskalos, A.M. de Oliveira Rev. Cienc. Vet. (Lisbon) 50:253:264, 1955.
13. Leite Velho, E. Bull. Off. int. Epiz. 46:335-340, 1956.
14. Scott, G.R. Bull. Off. int. Epiz. 63:645-677, 1965.
15. Heuschele, W.P. and Coggins, L. Bull. Epiz. Dis. Afr. 17:179-183, 1969.
16. Plowright, W., Parker, J. and Peirce, M.A. Vet. Rec. 85:668-674, 1969.
17. Heuschele, W.P. and Coggins, L. Proc. U.S. Livestock Sanit. Assoc. 69:94-100, 1965.
18. Sanchez Botija, C. and Badiola, C. Bull. Off. int. Epiz. 66:699-705, 1966.
19. Plowright, W. Comm. Eur. 5904 EN 575-587, 1977.
20. Walker, J. Thesis Univ. Zurich, Switzerland. London: Bailliere and Cox, 1933.
21. Pini, A. and Hurter, L.N. J. S. Afr. Vet. Assoc. 46:227-232, 1975.
22. Groocock, C.M. and Hess, W.R. Amer. J. Vet. Res. 44:591-594, 1980.
23. Sanchez Botija, C. Bull. Off. int. Epiz. 60:895-899, 1963.
24. Walton, G.A. Recent Adv. Acraology 11:491-500, 1979.
25. Hoogstraal, H. personal communication in: ASFV Brief Review. Arch. Virol. 76:73-90, 1983.
26. Plowright, W., Parker, J. and Peirce, M.A. Nature 221:1071-1072, 1969.
27. Plowright, W., Perry, C.T. and Greig, A. Res. Vet. Sci. 17:106-113, 1974.
28. Plowright, W., Perry, C.T. and Peirce, M.A. Res. Vet. Sci. 11:582-584, 1970.
29. Thomson, G.R., Gainaru, M.D. and Van Dellen, A.F. Onderstepoort J. Vet. Res. 47:19-22, 1980.
30. Thomson, G., Gainaru, M., Lewis, A., Biggs, H., Nevill, E., van der Pypekamp, H., Gerber, L., Esterhuysen, J., Bengis, R., Bezuidenhout, D. and Condy J. in: Proceedings of a CEC/FAO research seminar, Sassari, Sardinia, 23-25 Sept. 1981. EUR 8466 EN pp. 85-100, 1983.
31. Hess, W.R., Endris, R.G., Haslett, T.M., Monahan, T.J. and McCoy, J.P. Amer. J. Vet. Res. (in press).
32. Mellor, P.S. and Wilkinson, P.J. Res. Vet. Sci. (in press).

3
AFRICAN SWINE FEVER – CLINICAL ASPECTS

A. ORDAS ALVAREZ and M.A. MARCOTEGUI
Department of Animal Virology. Instituto Nacional de Investigaciones
Agrarias. Embajadores, 68 - 28012 Madrid, Spain.

African Swine Fever (ASF) is a disease which evolves under
different clinical forms, extensively described by many investiga-
tors, including MONTGOMERY (1921), WALKER (1933), DE KOCH et al.
(1940), DE TRAY (1957), MANSO RIBEIRO and ROSA AZEVEDO (1958 and
1961) and POLO JOVER and SANCHEZ BOTIJA (1961).

These authors make evident the close similarity between the
clinical symptoms of ASF, acute and subacute forms in particular,
and those of Classical Swine Fever (CSF). A remarkable difference
between them is that chronic forms are more frequent in ASF than in
CSF.

Due to the threat the disease represents to the world swine
production, a rapid diagnosis is essential to prevent diffusion in
a clean fully susceptible country.

Since the appearance of ASF in the Iberian Peninsula (Portugal
1957; Portugal and Spain, 1960), laboratory diagnostic techniques
have been improved and completed. But before,coming to these techniques, a good knowledge of the clinical symptoms is mandatory in
establishing a primary suspicion, specially in disease-free areas
up to that time. Experiences from the outbreaks occurred in France
(1964, 1968 and 1974), Malta (1978), Sardinia (1978), Cuba (1971,
1980), Brazil (1978), Dominican Republic and Haiti (1979) and final
ly Belgium (1985) confirm it. Ignorance of the clinical symptoms
and their similarity to those of CSF delay the first ASF diagnosis
in a country, since before suspecting the presence of ASF, pigs are
considered affected by CSF or any bacterial disease and only after
unsuccessful treatments to these diseases ASF is suspected, thus
favouring the initial spreading.

When ASF is present for long periods of time in a country, the
initial clinical forms of rapid evolution revert to slower clinical

courses and even new clinical forms appear.

In Spain, for example, since ASF appeared in 1960 to the present, it has been varying in its clinical aspects. At the beginning
the rate of mortality was practically 100%, modifying slowly along
with the epizooty, to find at the present outbreaks where mortality
is relatively low (2-3%) and, in some cases, inexistent.

Generally, subacute forms are most frequent in ASF. Naturally
occurring chronic forms are sparse since sick and suspected animals
are slaughtered once the positive diagnosis is confirmed, thus preventing studies of the disease's development under natural conditions.

Clinical pictures that ASF presents in domestic and Spanish
wild pigs (Sus scrofa ferus) are very similar.

INCUBATION PERIOD

In a complex disease, as ASF is, with its variable clinical
forms, determining the duration of the incubation period is very
difficult, since it may vary largely depending on a series of factors
including amount of virus, virulence, route of penetration, resistence
to the pig, natural or experimental infection, etc.

In the experimental infection, incubation period varies between
2 and 5 days when inoculated, and between 4 and 19 days when infected
by contact.

In the natural infection this period oscillates between 4 and
10 days, but sometimes it may last for 12 and even 20 days.

Nowadays in Spain, the most highly observed incubation period
in acute and subacute forms, which are the most frequent, lasts for
5 to 7 days.

In the incubation period and before fever appears, which is
the first symptom of the disease, the haemadsorption test of MALQUIST
demostrates virus in blood. This viremia may occur 24 hours before
the onset of fever thus representing an important factor for the
spreading of ASF before it is first diagnosed in a farm or in an
area free of the disease.

CLINICAL FORMS

ASF symptomatology is presently very similar to that of CSF,
a safe differentiation between clinical signs of both diseases is
therefore difficult to perform.

The clinical course of ASF may evolve under hyperacute, acute, subacute, chronic and subclinic· or inaparent forms.

Hyperacute form

It was formerly described by MONTGOMERY in 1921 and observed in Spain during the early years of the epizooty (1960, 1961 and 1962) when it was the most frequent in the field. Pigs died between 1 and 3 days after the manifestation of the first symptom. Generally, symptomatology is very scanty, death is usually in apoplectic form. There is a high temperature, between 41-42º C, accelerated breathing, hyperemia in skin, sometimes inappetence and sometimes they die eating.

Ineffectiveness and virulence of the virus in these cases are so high that mortality is practically 100%.

Acute form

This clinical form is most frequent in ASF. In Spain, from 1962 on, the clinical forms more often observed are acute and sub-acute.

After an incubation period which may vary between 4 and 6 days, the first clinic alteration observed is a rise of temperature up to 40-42º C, maintained until proximity of death, when it decreases to lower than normal. However, in some cases death takes place when temperature is at its maximum peak. This phase may pass unnoticed if thermic exploration is not performed, since the animal appears to be normal (Fig. 1).

During this period there is inappetence, capricious appetite, adynamia, conjunctivitis and, in some cases, ocular mucous or muco-purulent secretion.

Pigs affected are somnolent, stand with difficulty and some of them show uncoordinated movements. There are circulatory and vascular injuries characterized by cutaneous alterations with hyperemia (exantema) and cyanosis, specially at the ears, extremities and abdomen. These manifestations may present in form of violet irregular spots of variable size or in wide areas, hematomes and necrotic plaques.

In experimental cases, the study of the hematic formula reveals a marked leucopenia (linfopenia, monocytopenia and thrombo-cytopenia).

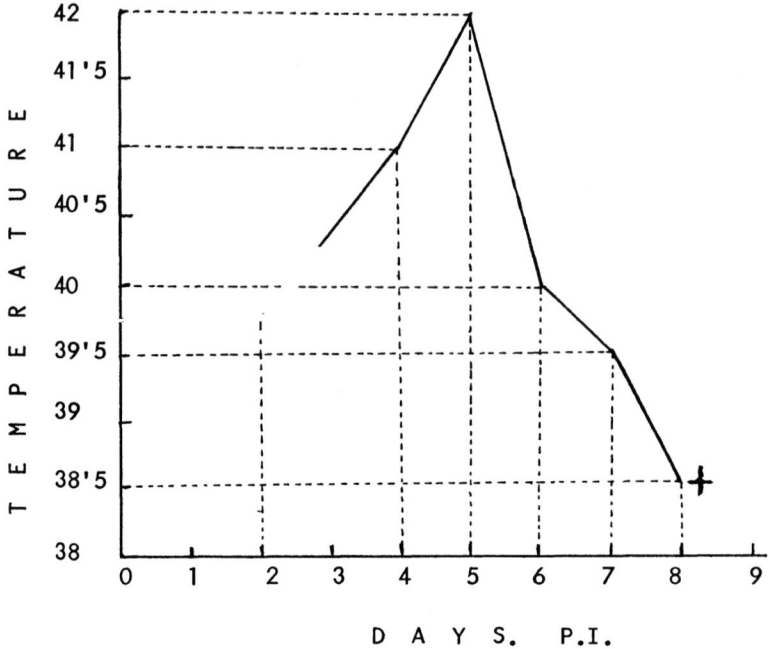

Fig. 1. Thermic curve. Acute form. + = death

If pregnant sows are in the farm, one of the first symptoms observed is miscarriage. This occurs generally at day 4-6 after febrils manifestations. Fetuses and fetal membranes show hemorragic lesions. Abortion is not caused by fever but as an effect of the virus, since in umbilical tissues significant a ounts of virus have been detected. Sometimes abortions are produced by secondary bacterial infections.

In the digestive tract the most persistent manifestation is constipation and vomiting, and less frequently diarrhoea. Eventually hemorragic discharges trough the anus are observed (melena).

Alterations of the respiratory apparatus show very early and are characterized by disnea, broken respiration and coughing. At nasal fossa levels, serous, sero-mucous or muco-purulent secretions may be detected due to pulmonary edema. In some cases, nasal hemo-rragies (epistaxis) may occur (Fig. 2).

As the disease advances, nervous disorders appear including

15

Fig. 2. Acute form. Epistaxis.

somnolence, unsteady walking, motor uncoordination, weakness of the
rear extremities, encephalitic signs, convulsive crises, falling to
the floor, and epistotones appear.

Death generally takes place between days 4-8 after appearance
of the first symptoms and is high. Pigs die by generalized circulatory
insufficience and cardiac failure.

A common observation in one farm is that at the beginning acute
forms are most frequent, its severity eventually subsiding; subacute
forms then appear, to end in chronic forms.

Subacute form:

Symptoms are very similar to those of the acute form, differing in intensity and duration of the disease.

After an incubation period of 6 to 12 days, pigs show a rise
in temperature and all symptoms of the acute form but, generally,
less accentuated (Fig. 3).

Mortality may vary depending on a series of factors as breed,
management, etc., but it is often high (60-90%). It takes place
between 6 to 10 days after the onset of symptoms and 12 to 25 days
after introduction of virus, due to a generalized lesion of the
reticulo-endothelial system which leads to vascular lesions determining
the cardiac failure.

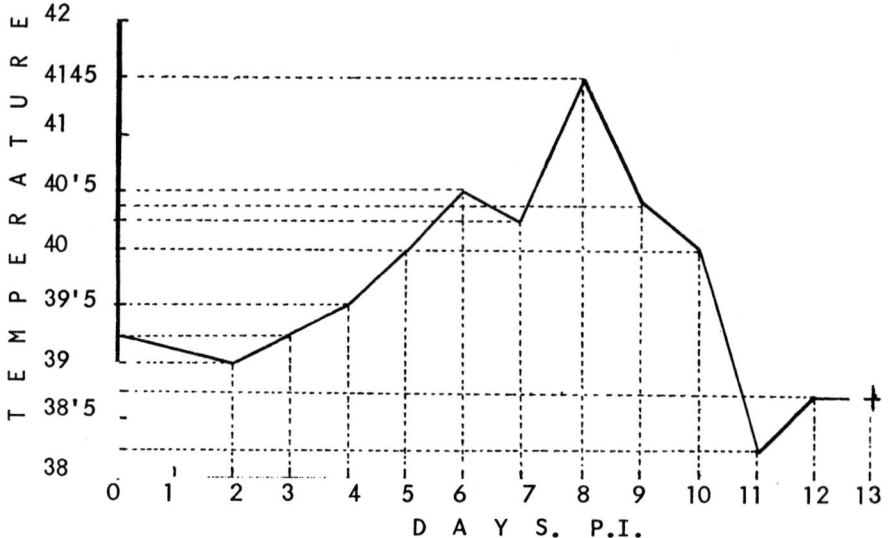

Fig. 3. Thermic curve. Subacute form. + = death.

Chronic form:

It is produced in those pigs which successfully survive the
subacute form, as described by DE KOCH et al. (1940), DE TRAY (1957)
and SANCHEZ BOTIJA (1962 and 1982).

Frequent symptoms are: discrete, 39,5-40,5º C., irregular and
ondulating fever, somnolence, capricious appetite, loss of weight
and delayed growth.

The clinical features mostly observed are those affecting
lungs, skin and joints.

Symptoms of the respiratory tract include pneumonia in all
its manifestations, dysnea, difficulty in breathing, etc. Sometimes
when the pig is submitted to a violent exercise, it may die by
asphyxia.

Skin lesions are characterized by formation of plaques or
nodules (Fig. 4), ulcerations with necrosis and loss of substance
at the ears, joints, tail and snout (Fig. 5). Some skin lesions are
similar to those produced by allergic phenomena affecting wide

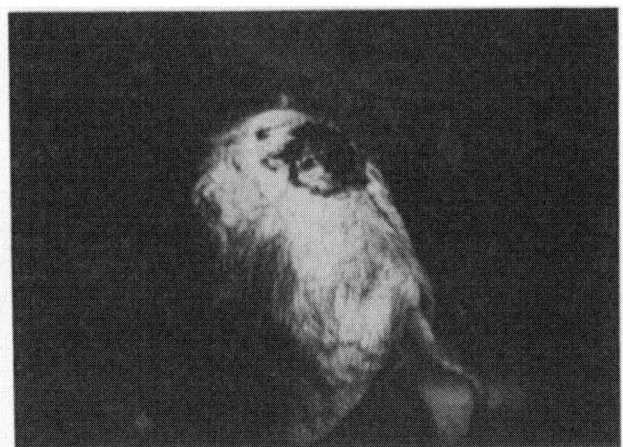

Fig. 4. Chronic form. Ulceration with necrosis of skin.

extensions of the back and shoulders; sometimes subcutaneous abcessés are produced by immunopathologic reactions.

Soft, painless swelling over the leg joints may be observed (multiple arthritis), more frequently in carpal and tarsal joints, and also over the jaw (Fig. 6).

The chronic forms' course lasts from one to several months, alternating activation and remission of symptoms.

Mortality is highly variable and among its causes are pneunonia, cardiac failure, hyalinization of the muscle fibers on the miocardium. It only occurs in some pigs (dripping death), the rest usually recover, death sometimes occurs after an activation of the disease with acute ASF clinical picture.

Subclinic or inapparent form

This ASF clinical form was observed primarily by MONTGOMERY (1921) in one domestic pig exposed to the infection by contact and later by DE TRAY (1957) although it is not frequent among African domestic pigs. Nevertheless, this form is current among wild pigs in Africa (Phacochoerus, Potamochoerus and Hylochoerus).

In Spain, this clinical forms were not found in the early years of the epizooty and only after 3-4 years they were detected, rising in number with the passage of the time (SANCHEZ BOTIJA, 1963).

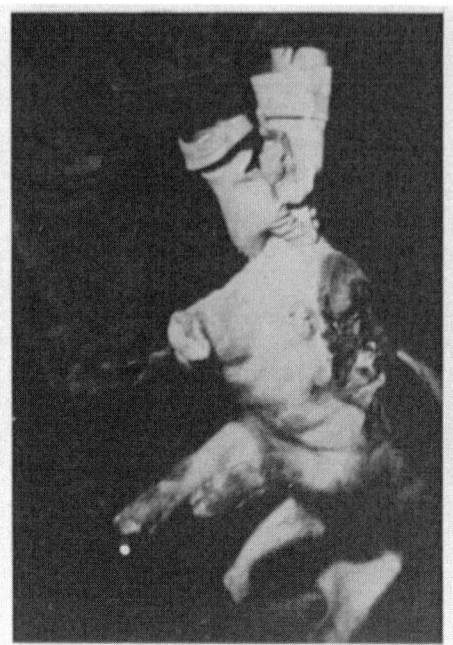

Fig. 5. Chronic form. Loss of the ear due to
ulceration and further necrosis

These subclinical cases are due to infections with field virus
strains of low virulence. Infected pigs show a discrete disease with
slow course, with no clinically noticiable functional alterations
and appear to be normal.

Subclinical status may also derive from pigs which have recover
ed from an ASF subacute episode and appear to be normal. Nevertheless
these pigs are virus carriers and at necropsy show discrete lesions.
These forms are not clinically detectable and demonstration of antibodies in serum by means of laboratory techniques (indirect immunofluorescence, ELISA; etc.) is necessary.

The increase of these clinical forms in an infected country
brings about a continuous sanitary surveillance to detect these
carrier pigs, which may give rise to hidden outbreaks that may prove
difficult to control and eradicate ASF.

19

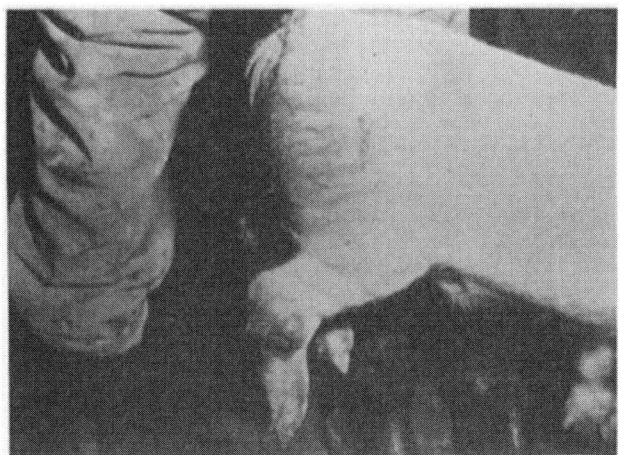

Fig. 6. Chronic form. Soft, painless swelling in leg joint.

Atypical form

The so called atypical forms are just chronic forms originated by vaccine virus strains which show residual virulence. They appear several months after the use of a vaccine virus.

Primarily they were thought to be due to microbial complications (bacteria and mycoplasma) but it was later demonstrated that they were originated by immunopathologic phenomena produced by antigen-antibody reaction in presence of complement.

These clinical forms generally present a picture very similar to that of the chronic form with pulmonary, cutaneous and articular symptomatology.

REFERENCES

1. De Tray (1957). African Swine Fever: a review. Bull. Dis. Afr., 5: 475-478
2. Lucas, A., Haag, J. et Larenaudy, B.C. (1970). Peste Porcine Africaine (Maladie de Montgomery). Ed. L'Expansion Scientifique Française. Paris
3. Manso Ribeiro, J., Nunes Petisca, J.L., Lopes Trazao, F. et Sobral, M. (1963). Vaccination contre la Peste Porcine Africaine. Bull. Off. Int. Epiz., 60: 921-937.
4. Neitz, W.O. (1964). Enfermedades de importancia naciente de los animales. Ed. FAO: Estudios agropecuarios, 1-74.
5. Polo Jover, F. y Sánchez Botija, C. (1961). Peste Porcina Africana en España. Bull. Off. Int. Epiz. 55: 107-147:

6. Sánchez Botija, C. (1962). Estudios sobre la Peste Porcina Africana en España. Bull. Off. Int. Epiz., 58: 707-727.

7. Sánchez Botija, C. (1965). Present characteristics of ASF in Spain. FAO/OIE International Meeting on Hog Cholera and African Swine Fever, 31 May - 5 June. Rome.

8. Sánchez Botija, C. y Ordás, A. (1980). Peste Porcina Africana. Patología y clínica del ganado porcino. Ed. Noticias Neosan, 119-154.

9. Sánchez Botija, C. (1982). Peste Porcina Africana: nuevos desarrollos. Rev. Sci. Tech. Off. Int. Epiz. 1 (4): 1065-1094.

10. Schlaffer, D.H. and Mebus, C.A. (1984). Abortion in sows experimentally infected with African Swine Fever virus: Clinical features. Am. J. Vet. Res. Vol 45, 7: 1353-1360.

4

PATHOBIOLOGY AND PATHOGENESIS

CHARLES A. MEBUS

USDA, ARS, PIADC, P.O. Box 848, Greenport, NY 11944 U.S.A.

ABSTRACT
     Histologic lesions in pigs induced by a highly virulent African
swine fever virus (Lisbon 60) were compared with lesions induced by a
moderately virulent ASF virus (Dominican Republic). The location of
lesions in both groups of pigs suggest that ASF virus replicates in the
mononuclear-phagocytic system and has a predilection for specific
lineages of reticular cells (antigen processing cells). The histologic
differences in the effect of high and moderately virulent ASF virus
infection on these antigen processing cells suggests that the fate of
these cells is a major factor in the development of an immune response
and survival or death of the pig.

INTRODUCTION
     African swine fever virus (ASF) is immunologically an interesting
disease. Pigs infected with a highly virulent virus i.e. Lisbon 60 (L60),
so designated because 100 % mortality can be expected 6-10 days after
inoculation, develop little or no circulating antibody. Pigs infected
with a moderately virulent virus i.e. Dominican Republic (DR) isolate, an
isolate that causes little or no mortality, develops antibody 4-6 days
after infection. This antibody does not neutralize the virus or clear
the viremia but does block direct immunofluorescence (1). The antibody
response is associated with temperature and leucocyte counts gradually
returning to normal and the viremia gradually decreasing to below detectable levels about 35 days after infection (2). The difference in the
immune response in pigs infected by high or moderately virulent ASF isolates could be explained by infection of lymphocytes (3,4 and 5) or by
infection of macrophages and reticular cells (6). The purpose of this
study was to compare the histologic lesions induced by a highly virulent
and a moderately virulent ASF virus to determine if viral virulence was

related to histopathology and to determine if histopathological differences were related to the development of the immune response.

MATERIALS AND METHODS
Virus.

The highly virulent virus was the L60 isolate and the moderately
virulent virus was a DR isolate. The inoculum for both isolates was
homogenized spleen and liver diluted in phosphate buffered saline. The
L60 inoculum had a titer of $10^{5.8}$ hemadsorbing doses $(HAD)_{50}/ml$ and the
DR inoculum a titer of $10^{4.7}$ HAD_{50}/ml. The pigs were inoculated
intranasally-orally (INO) with 4ml. of the respective inoculum.

Pigs.

Four groups of 18 to 28 kilogram pigs were housed in separate
isolation rooms. The L60 inoculated pigs consisted of one group of five
pigs which were killed or died 4, 7, and 8 days post inoculation (DPI)
and a second group of four which were killed or died on 8 and 9 DPI.
The DR inoculated pigs consisted of one group of five pigs, one of which
was killed at 4, 7, 8, 13, and 16 DPI and a group of 13 pigs, two of
which were killed on days 4, 6, 8, 10, 12, and one at 14 DPI.

At necropsy, tissues were collected for culture and histologic
examination. Tissues for virus isolation were stored at -70C. Tissues
for light microscopy were fixed in 10% phosphate buffered formalin and
those for immunofluorescent microscopy were imbedded in O.C.T. Compound
(Miles Laboratory) and stored at -70C. Fixed tissues were embedded in
paraffin, sectioned and stained with haemotoxylin and eosin. Frozen
tissues were sectioned and stained with fluorescein conjugated porcine
ASF DR antibody.

Virus Titration.

Virus titrations were done using porcine buffy coat cultures.
Cultures were inoculated with 10-fold dilutions of blood or homogenized
tissue supernatant and examined daily for hemadsorption.

Antibody detection.

Serum antibody to ASF virus was detected by indirect immunofluorescence using ASF virus infected Vero cells on coverslips and
fluorescein conjugated sheep anti-swine IgG.

RESULTS

The incubation periods and onset of fever, for both the high and moderately virulent viruses were similar. The animals were viremic by 2 DPI and febrile by 2-4 DPI.

Gross lesions.

The L60 infected pigs euthanatized at 4 DPI had no gross lesion. Pigs which were euthanatized or died 7-9 DPI had lesions similar to those described for classical AFS: reddened areas of skin, enlarged submandibular lymph nodes, markedly enlarged dark friable spleen, enlarged and severly hemorrhagic gastrohepatic and renal lymph nodes, petechial hemorrhages in the kidneys, and edema of the lungs.

The DR isolate infected pigs euthanatized 4-10 DPI had mild hemorrhage in the gastrophepatic and renal lymph nodes; the spleens were mildly enlarged but had a normal color and texture.

Viral titers

Viremia titers in the DR isolate infected pigs ranged from $10^{5.6}$ to $10^{7.3}$ HAD_{50}/ml. Viral titers in gastrohepatic lymph nodes and spleens ranged from 10^5 to 10^6 and 10^6 to 10^7 HAD_{50}/gm respectively.

Immunofluoresence.

Spleen, lymph nodes, lung, liver, and kidney from all L60 isolate infected pigs which were euthanatized or died at 4-9 DPI were positive for ASF antigen by direct immunofluoresence.

Spleen, lymph nodes, lung, liver, and kidney from DR isolate infected pigs euthanatized at 4 DPI were positive for ASF antigen by direct immunofluoresence. These same tissues from pigs euthanatized at 6 and more DPI were negative for ASF antigen by direct immunofluoresence. Sera from pigs negative by direct immunofluoresence were positive by indirect immunofluoresence for ASF antibody.

Histologic lesions in highly virulent ASF virus infected pigs.

Histologic lesions in lymph nodes, spleens, livers, tonsils, and brains from pigs that were euthanatized or died at 4-9 DPI with highly virulent ASF virus (L60) were as follows:

Inguinal lymph nodes.

At 4 DPI, the nodes were normal. At 7 DPI, there was mild peripheral hemorrhage and postcapillary venules had thickened eosinophilic walls, enlarged endothelial nuclei and an occasional pyknotic endothelial cell nucleus. At 8 DPI, 1 of 4 pigs had vascular changes similar to the 7 DPI pigs; 2 pigs had pyknotic nuclei in the inter-follicular areas; and 1 pig had normal nodes. At 9 DPI, 1 of 2 pigs had many pyknotic nuclei in the cortico-medullary and interfollicular areas; the other pig had normal nodes.

Submandibular lymph nodes.

At 4 DPI, the nodes were normal. At 7 DPI, there were numerous pyknotic nuclei in the interfollicular areas; postcapillary venules had thickened homogeneous eosinophilic walls and many mononuclear cells in the lumens. Nodes from 8 and 9 DPI pigs had a decrease in lymphocytes, pyknotic nuclei in germinal centers, and postcapillary venules with thickened eosinophilic walls. In addition, 3 of the 6 pigs had petechial hemorrhages associated with the necrosis. Numerous necrotic cells were concentrated in the paracortical areas and extended into the interfollicular areas. Two pigs had extensive areas of necrosis and some hemorrhage.

Mesenteric lymph nodes.

At 4 DPI, the nodes were essentially normal. At 7 DPI, reticular cells in the interfollicular area had abundant eosinophilic cytoplasm and some postcapillary venules had thickened eosinophilic walls. At 8 DPI, nodes from 3 pigs had small areas of hemorrhage and the nodes from 1 pig also had pyknotic nuclei in the paracortical areas and thickened postcapillary venules. At 9 DPI, the nodes had areas of hemorrhage; cells in the paracortical areas were enlarged and some had mitotic figures; pyknotic nuclei in the interfollicular areas varied from scat-tered to numerous; and postcapillary venules had thickened walls.

Gastrohepatic lymph nodes.

At 4 DPI, the nodes had scattered small areas of edema, hemorrhage, enlarged reticular cells and scattered pyknotic nuclei in the inter-follicular areas. At 7 DPI, there were pyknotic nuclei and hemorrhage in the interfollicular areas, depletion of lymphocytes, and a few pykno-tic nuclei in germinal centers. The nodes from 3 of 4 pigs killed at 8 DPI had extensive hemorrhage in the medullary areas and pyknotic nuclei in the interfollicular areas. At 9 DPI, the nodes had large amounts of fluid and/or hemorrhage and pyknotic nuclei in the medullary areas.

Spleen.

At 4 DPI, the marginal zone of the periarterial lymphoid sheaths had scattered concentrations of erythrocytes, numerous pyknotic and fragmenting nuclei, and reticular cells with increased amounts of cytoplasm. At 7-9 DPI, the red pulp was congested, periarterial macrophage sheaths were necrotic, and periarterial lymphoid sheaths had pyknotic nuclei and were depleted of lymphocytes.

Liver.

At 4 DPI, the liver was normal. At 6-9 DPI, there were scattered pyknotic nuclei in the lymphoreticular cell infiltrate in the interlobular and portal areas, widely scattered pyknotic nuclei in the sinusiods, and widened spaces of Disse. In addition, one 8 DPI pig had foci of congestion in which the severity of hepatocyte necrosis involvement varied from scattered individual cells to necrosis of complete hepatic cords and to necrosis of groups of hepatic cords.

Tonsils.

Tonsils from pigs 7 and 8 DPI had a depletion of lymphocytes, and pyknotic nuclei in interfollicular areas.

Brain.

At 4 DPI the brain was normal. Brains from pigs at 7-9 DPI had a mild lymphoreticular cell cuffing of scattered vessels and and a mild lymphoreticular cell infiltrate and a few pyknotic nuclei in the choroid plexus of the lateral ventricle. Three pigs also had a mild nonsuppurative meningitis.

Histologic lesions in moderately virulent ASF virus infected pigs.

Histologic lesions in lymph nodes, spleens, livers, tonsils, and brains from pigs that were euthanatized or died 4-16 DPI with the moderately virulent ASF virus (DR) were as follows:

Inguinal lymph node.

The nodes from a 4 DPI pig contained many eosinophils. Nodes from a 7 DPI pig in addition to the eosinophils had post capillary venules with eosinophilic thickened walls and enlarged endothelial cells. Nodes from a 13 DPI pig had a few post capillary venules with thickened walls and enlarged endothelial cell nuclei. An 8 and 16 DPI pig had normal nodes.

Submandibular lymph nodes.

At 4 DPI, nodes from 1 of 3 pigs were normal. Those from a second pig were congested and those from a third pig had several foci of pyknotic nuclei in interfollicular areas and enlarged endothelial cells. At 6 DPI, nodes from 1 of 2 pigs were normal; the other pig had few pyknotic nuclei and mitotic figures in the germinal centers. Nodes from 7 DPI pigs had cortical areas composed of fairly uniform reticular cells in which there were quite a few mitotic figures and the post capillary venules had thickened eosinophilic walls and enlarged endothelial cell nuclei. At 8 DPI, the nodes from 1 of 3 pigs had a few mitotic figures. Those from a second pig had hemorrhagic areas in which there were numerous pyknotic nuclei. Nodes from the third pig were hemorrhagic and a few germinal centers had necrotic cells. Nodes from 10, 12 and 13 DPI pigs were similar to 8 DPI nodes. Nodes from 14 and 16 DPI pigs were essentially normal.

Mesenteric lymph nodes.

The nodes from a 4 and 8 DPI pig were normal. Nodes from an 8 and 13 DPI had enlarged reticular cells and an occasional mitotic figure in the interfollicular areas. A 16 DPI pig had enlarged reticular cells.

Gastrohepatic lymph nodes.

At 4 DPI, nodes from 2 pigs had mild peripheral hemorrhage and 1 pig also had scattered pyknotic nuclei in the interfollicular areas. At 6 DPI, nodes from 2 pigs had mild to moderate peripheral hemorrhage, scattered pyknotic nuclei in hemorrhagic areas, mitotic figures in interfollicular areas and post capillary venules with thickened eosinophilic walls and enlarged endothelial cell nuclei. Nodes from pigs killed at 8, 10, 12, 13, and 16 DPI were similar to the nodes from the 6 DPI pigs.

Spleen.

At 4 DPI, the spleen from 1 of 2 pigs had a few scattered pyknotic nuclei in the red pulp and the second pig had pyknotic nuclei in several periarterial macrophage sheaths. At 6 DPI, the spleen from 1 of 2 pigs had possible enlargement of reticular cell nuclei in the red pulp; the second pig had pyknotic nuclei in the periarterial lymphoid sheaths, more pronounced eosinophilic staining of the periarterial macrophage sheaths and the areas of erythropoiesis. Spleens from pig 12, 13, and 16 DPI had erythropoiesis. The spleen from a 14 DPI pig was normal.

Liver.

Enlarged Kupffer cells in a 7 DPI liver was the only change associated with ASF infection.

Tonsils.

At 6 and 7 DPI there were a few mitotic figures in the interfollicular areas.

Brain.

One pig at 6 DPI had a mild lymphoreticular cell infiltrate in the
choroid plexus. Two 11 DPI pigs had mild lymphoreticular cell infiltrate
in the choroid plexus, meninges and Virchow-Robbins space of a few
vessels. Brains from the other 14 pigs were normal.

DISCUSSION

The three most striking differences in the infections induced by
the highly virulent and the moderately virulent ASF viruses were: 1.
Very high mortality in pigs infected with the highly virulent virus
versus a low mortality in pigs infected with moderately virulent ASF
virus; 2. Extensive necrosis especially in the spleen and certain lymph
nodes from pigs infected with the highly virulent virus versus little
necrosis in the moderately virulent infection; 3. No or little antibody
production in the highly virulent infection versus an essentially normal
antibody response in moderately virulent infections. Replication of ASF
virus in monocytes and macrophages is well documented (7,8). In
contrast to hog cholera (classical swine fever), ASF virus does not
infect epithelial cells. ASF virus, from the location of lesions
reported here and in the literature (9,10), apparently has a predilection
for the red pulp, marginal zones of the periarterial lymphoid sheaths,
and periarterial macrophage sheaths of the spleen; the paracortical
areas, interfollicular areas and post capillary venules of the lymph
nodes; and Kupffer cells of the liver. Immunologists report localization of ingested foreign material and antigen in the periarterial
macrophage sheaths, red pulp, and marginal zones of the periarterial
lymphoid sheaths in the spleen, and medullary areas and follicular
marginal zones of the lymph nodes (11). These sites have in common phagocytic cells of the reticulo-endothelial system. There is evidence for
the involvement of follicular dendritic cells (12), (dendritic reticular
cells), macrophages (13), Langerhan cells and Kupffer cells in presen-

tation of antigen to T and B lymphocytes for development of an immune
response. Therefore, it is proposed that ASF virus replicates in cells
which process antigen. Evidence against ASF virus directly infecting B
or T lymphocytes is that in fatal infections lymphocytes in the
follicles of lymph nodes (B cells) and lymphocytes in the splenic
periarterial lymphoid sheaths (T cells) are still present after there is
extensive necrosis in the interfollicular areas of lymph nodes and
splenic marginal zones.

Initial infection by ASF virus causes stimulation of cells of the
reticulo-endothelial system, reported here and in the literature (4,10),
as evidenced by increased cell size and mitosis. In infection by highly
virulent ASF virus, it appears that the takeover of the cell is so rapid
and complete that the cells die before stimulation is too evident.
Stimulation is best seen in infections with moderately virulent ASF
virus; the cells and nuclei are large and mitotic figures are not uncommon. This finding suggests that less virulent ASF viruses have attained
a commensal relationship with the cell, because even though the virus
replicates to high titer, sufficient normal cell metabolism must persist
for most cells to survive. These surviving cells apparently retain the
ability to process antigen because anti-ASF IgM was present in detectable
amounts in serum at 4 DPI (J. M. Sanchez-Vizcaino, INIA, Madrid, Spain,
personal communication) and anti-ASF IgG was found at 6 DPI in the DR ASF
infected pigs but not in L60 infected pigs. The antibody response in
moderately virulent ASF virus infection thus appears to be similar to
conventional antigenic stimulation. However, in a conventional viral
disease in which virus neutralizing antibody is produced, the presence of
detectable antibody coincides with the inability to detect viremia. This
is not the case in ASF infection because the antibody produced does not
neutralize the virus. Development of ASF antibody is frequently followed
by an increasing peripheral leucocyte count and decreasing fever; thus,
it appears that the effect of ASF virus infection on reticular cells and
development of an immune response is related to whether or not the pig
survives an ASF infection. How this immune response operates still has
to be determined.

REFERENCES

1. Heuschele, W.P., Coggins, L., and Stone, S.S.: (1966). Am. J. Vet. Res., 27, 477-484, 1966.
2. Mebus, C.A. and Dardiri, A.H.: Pro. U. S. Animal Health Assoc. Richmond, Virginia, Carter Composition Corp. 1979, pp. 227-239.
3. Kono, S., Taylor, W.D., Hess, W.R., and Heuschele, W.P.: Cornell Vet., 62 486-506, 1972.
4. Nunes Petisca, J.L. and Martins Goncalves, J.M.: Hog Cholera/Classical Swine Fever and African Swine Fever (EUR 5904 EN). Luxembourg, Commission of the European Communities, 1977, pp. 612-627.
5. Wardley, R.C. and Wilkinson, P.J.: Res. Vet. Sci., 28, 185-189, 1980.
6. Moulton, J. and Coggins, L.:Cornell Vet., 58, 364-388, 1968.
7. Hess, W.R. and DeTray, D.E.: Bull. Epiz. Dis. Africa, 8, 317-320, 1960.
8. Wardley, R.C. and Wilkinson, P.J.: J. Gen. Virol., 38, 183-186, 1978.
9. De Kock, G., Robinson, E.M. and Keppel, J.J.G.: Onderstepoort J. Vet. Sci. Anim. Ind., 14, 31-93, 1940.
10. Kono, S., Taylor, W.D., Dardiri, A.H.: Cornell Vet., 61, 71-84, 1971.
11. Bach, J.F. (ed). In Immunology, Wiley Medical Publications, New York, 1978, pp. 28-30.
12. Klaus, G.G.B., Humphrey, J.H. Kunkl, A. and Dongworth, D.W.: Immunological Review 53, 3-28, 1980.
13. Schrader, J. W. and Nossal, G. J. V.: Immunological review 53, 61-85, 1980.

5

MOLECULAR BIOLOGY OF AFRICAN SWINE FEVER VIRUS

ELADIO VIÑUELA
Centro de Biología Molecular (CSIC-UAM), Facultad de Ciencias, Universidad Autónoma, Canto Blanco, 28049 Madrid, Spain.

ABSTRACT

African swine fever virus particles have a size of about 200 nm and a lipoprotein envelope surrounding an 80 nm virus core.the virus contains 30-40 proteins,which include many enzymes able to synthesize,cap,methylate and polyadenylate functional messenger RNA.The African swine fever virus genome is a noninfectious double-stranded DNA molecule of about 170,000 base pairs. At both termini the two DNA strands are covalently linked by a partially base-paired hairpin loop into a single polynucleotide chain.the virus particle can synthesize in vivo functional virus mRNAs in the absence of protein biosynthesis. One of the major biological problems posed by African swine fever virus-infected animals is that the virus-specific antibodies are not neutralizing. A possibility that may account for this atypical immune response is antigenic variability,that has been made clear by a comparison of the binding properties of a collection of monoclonal antibodies to different African swine fever virus field isolates. The presence in African swine fever virus DNA of several multigene families, a finding not reported for any other virus, may be related to the ability of African swine fever virus to evade the immune system.

INTRODUCTION

African swine fever (ASF) virus is a menace to the pig population in the world because there is no vaccine. The virus propagates in ticks and changes easily and different virus isolates produce diseases with different clinical symptoms or no disease at all.the control and eradication of ASF requires rapid diagnosis, drastic slaughter and quarantine (1-3).

ASF was first described in 1921 by Montgomery, who reported several disease outbreaks of domestic pigs in Kenya since
1910 with a mortality close to 100% (4). In 1957 the disease
appeared outside Africa, in Portugal, and in 1960 spread to
Spain. In the 1960s and 70s it was found in France, Italy,
Sardinia and Malta and, in America,in Cuba, Brazil, Dominican
Republic and Haiti. In early 1985 there was an ASF outbreak
in Belgium and in 1986 in The Netherlands. Today, the disease
is enzootic in Sub-Saharan Africa, Portugal, Spain, Sardinia,
Brazil and Haiti.

ASF is produced by an icosahedral deoxyvirus of about 200
nm which infects only species of the Suidae family and ticks
of the genus Ornithodoros (family Argasidae). It is the only
known arbovirus which contains DNA. All attempts to infect
other animal species with ASF virus in an attempt to find a
more adequate experimental model have been unsuccessful (5).
In pig tissues, viral antigens have been associated with macrophages and reticular cells, whereas in blood and bone marrow
the main cells involved in the infection are monocytes, polymorphs and megakaryocytes. Cells lining some blood vessels
degenerate only late in infection (6).

The ASF virus-specific antibodies induced in the infected
animals are neither protective in vivo nor neutralizing in
vitro (4,7). The reasons for this peculiar immune response are
unknown. It is likely that, without knowledge of the ASF virus
evasion mechanism from the host immune system, it will be difficult to obtain a vaccine. For this, it will be necessary to
have a better understanding of the structure and components of
the virion and the details of infection of sensitive cells.

ASF virus replicates in vitro in pig monocytes and macrophages, a small fraction of polymorphs and, perhaps, some
endothelial cells, but not in resting, or mitogen-stimulated
T or B lymphocytes (8-11).Macrophages from resistant animals
do not produce infectious virus in vitro (12). After adaptation, some ASF virus isolates multiply in primary cell cultures from pigs or in established cell lines from swine or virus-
resistant animal species (3). Most of the studies mentioned in

this review have been carried out with the virus isolate BA71, adapted to VERO cells, BA71-V (13).

VIRUS PURIFICATION – STRUCTURAL PROTEINS.

Extracellular ASF virus particles, essentially free of membranes, can be obtained by equilibrium centrifugation in Percoll density gradients. The purified virus sediments as a single component in sucrose velocity gradients with a sedimentation coefficient of 3,500±300S, has a DNA-protein ratio of 0.05±0.01 and a specific infectivity of about 5 x 10^7 plaque-forming units/μg of protein. The total to infectious particle ratios in purified preparations is 10-50. The virus infectivity does not decrease after storage at -70ºC for at least 7 months (14).

Polyacrylamide gradient gel electrophoresis of virus labeled with either ^{35}S-methionine or ^{14}C-amino acids and dissociated with sodium dodecylsulphate and mercaptoethanol separates about 34 bands. The molecular weight values of the virus proteins range from 150 to 10 K (1 K = 1000 daltons) (Fig.1). A similar profile is obtained when the proteins are stained with silver nitrate. Some of these proteins are enzymes, involved in the synthesis in vitro of the viral mRNAs that are indistinguishable from those synthesized in the infected cells in the presence of cytosine arabinoside, an inhibitor of the viral DNA synthesis (Table 1).

Up to one-third of the ASF virus structural proteins might be cellular proteins, since the corresponding bands are more intense in virus particles purified from cells labeled with ^{35}S-methionine before infection than in those purified from cells labeled after infection. These proteins might account for the reactivity of purified ASF virus with antisera raised against uninfected cells, and some, if not all, of them are incorporated into the virions since the infectious particles can be precipitated with sera against uninfected cells in the presence of Staphylococcus aureus (14). The biological significance of the presence of these cellular components in the virus particles, specially their possible relationship

34

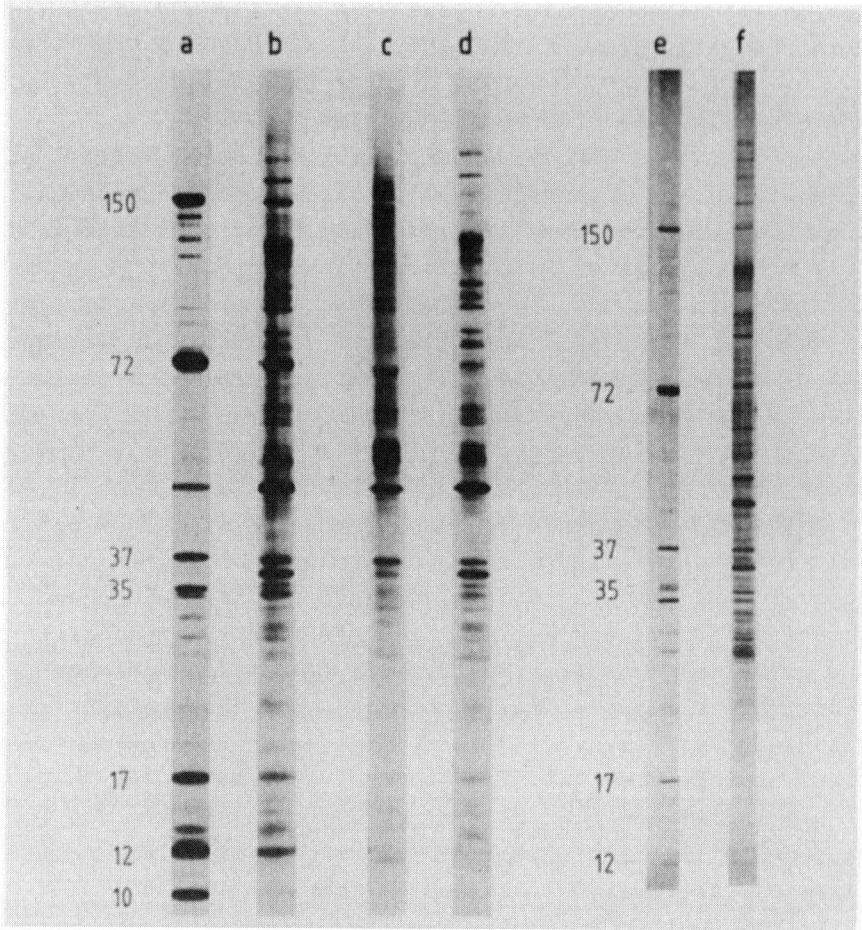

Fig. 1. Structural proteins of extracellular ASF virus particles. Polyacrylamide gel electrophoresis of Percoll-purified
ASF virus (a,c,and e) and vesicles (b,d,and f) obtained from
unlabeled cells (e and f) and from cells labeled with 35S-
methionine after (a and b) or before (c and d) infection. An
equal amount of label was placed in each lane. The specific
radioactivities of the samples were 7,541, 11,631, 2,143 and
9,657 cpm/µg of protein in lanes a, b, c and d, respectively.
Radioactive and unlabeled protein bands were detected by fluo-
rography and silver staining, respectively. Lanes a through d
and lanes e and f were from two different gels. Molecular weights
(in thousands) are indicated. Reprinted with permission from (14).

Table 1. Enzymatic activities present in ASF virus particles.

RNA polymerase

$nNTP \rightarrow RNA + nPP_i$

PolyA polymerase

$nATP + N(-N)_m \rightarrow N(-N)_n (-A)_n + nPP_i$

Nucleoside triphosphatase

$(d)NTP + H_2O \rightarrow (d)NdP + P_i$

Stimulated by RNA or DNA

RNA guanylyltransferase

$GTP + ppN(-N)_m \rightarrow GpppN(-N)_m$

RNA (guanine-7-) methyltransferase

$AdoMet + GpppN(-N)_m \rightarrow m^7GpppN(-N)_m + AdoHcy$

RNA (nucleoside-2'-) methyltransferase

$AdoMet + m^7GpppN(-N)_m \rightarrow m^7GpppN^m(-N)_m + AdoHcy$

Deoxyribonuclease (pH 4.5 optimum)

$DNA + H_2O \rightarrow$

Deoxyribonuclease (pH 7.5 optimum)

$DNA + H_2O -$

DNA topoisomerase

Coumermycin sensitive

Protein kinase

$ATP + Protein \rightarrow P\text{-}Protein + ADP$

---

with the peculiar ASF virus-induced immune response, is unknown.

In contrast to all the viruses with a lipoprotein envelope, ASF virus lacks glycoproteins, a property that it
shares with the iridoviruses. However, highly purified ASF
virus particles contain in their surface two non-protein,
glycosylated components of apparent molecular weight 230 and
90 K that interact specifically with some lectins (15).

VIRUS STRUCTURE

The ASF virus morphology is similar to that of the iridovirus that infect vertebrates. The particle consists of a
core, with a diameter of about 80 nm, that contains the DNA,
a lipoprotein envelope surrounding the virus core and the
capsid. The extracellular virus has an external envelope,
derived by budding through the plasma membrane (16).The capsid
consists of a hexagonal arrangement of capsomeres which appear
as 13 nm long hexagonal prisms, each with a central hole. The
intercapsomere distance is about 8 nm and the triangulation
number equals or is larger than 208 (17) (Fig. 2).

The critical antigens that induce the synthesis of either
protective antibodies or cytotoxic T lymphocytes against infectious agents are either external proteins in the virion or
virus-induced proteins incorporated in the infected cell membrane, respectively. The identification of these antigens is
important to study the immune response of the infected animals.
The use of monoclonal antibodies specific for different ASF
virus structural proteins (18) has allowed identification by
immunoelectron microscopy of some antigens, such as proteins p14,
p12, p24, that map at the outer region of the virus, among
others (19). Treatment of highly purified ASF virus particles
with either non-ionic detergents or ^{125}I has allowed identification
of proteins p35, p17, p14, p12 and p10 in the virion periphery.

DNA

The ASF virus genome is a large double-stranded DNA
molecule of about 170 kb (1 kb = 1000 nucleotide pairs). In

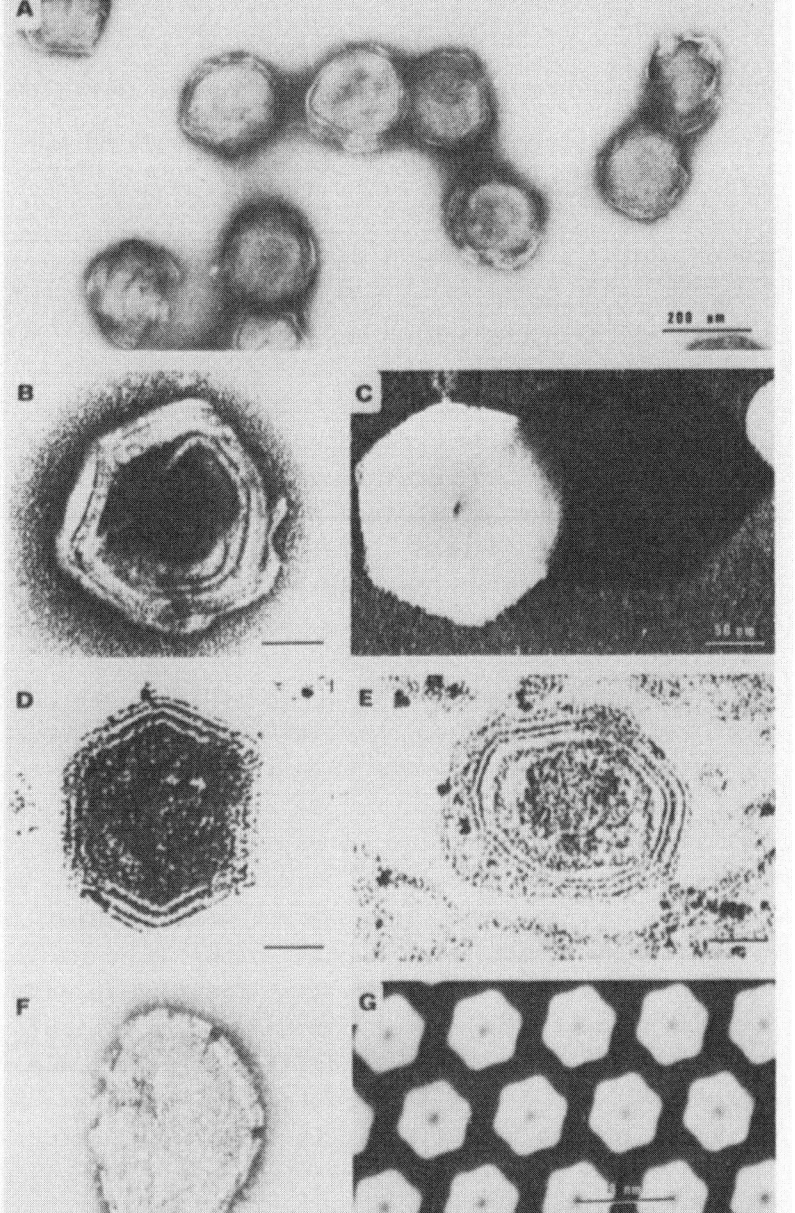

Fig. 2. Structure of ASF virus particles. A. Purified virus
fixed with glutaraldehyde and negatively stained with sodium
phosphotungstate; the particles are not permeable to the contrasting agent and show few structural details. B. In a broken
particle the penetration of the contrasting agent allows the virus
layers to be seen.   C. A freeze-dried particle, shadowed
with platinum; the hexagonal outline of virus and shadows shows
the icosahedral symmetry of the virion. D, E. Thin-sections
of well (D) and badly (E) preserved particles. F. Negatively
stained capsid of virus treated with detergent to remove the
external membrane, showing the periodic structure of the capsid. G. Tridimensional reconstruction of the capsids shown
in F. Reprinted with permission of (17).

contrast to a previous claim (20), ASF virus DNA is not infectious, consistent with the presence within the virion of the
enzymes necessary for DNA transcription (21,22). At both ends
the two DNA strands are covalently linked by a partially
base-paired hairpin loop into a single polynucleotide chain
with a length of 37 nucleotides, composed almost entirely of
A and T residues. The loops at each DNA end are present in
two equimolar forms that, when compared in opposite polarities,
are inverted and complementary (flip-flop) as in the case of
poxvirus DNA. The hairpin loops of ASF and vaccinia viruses
have no homology, but both DNAs have a 16 nucleotide-long
sequence, close to the hairpin loops, with a homology of
about 80 percent. Following the hairpin loops there is a 2440
nucleotide-long perfect terminal inverted repeat which
consists of unique sequences interspersed with 42 direct repeats
in tandem of 34 nucleotides, 5 repeats of 24 and 3 of 33 nucleotides. Some regions in the terminal inverted repeats of
ASF DNA show up to an 80 percent homology with similar regions in vaccinia virus DNA (A. González, V. Calvo and E.
Viñuela, unpublished data). A comparison of the genome of
several ASF virus isolates by restriction endonuclease map-
ping has shown the existence in ASF virus DNA of a constant
central region of 125 kb and two variable regions located
within and near the terminal inverted repeats (R. Blasco,
M. Agüero and E. Viñuela, unpublished data). Most of the
properties of ASF virus DNA are similar to those of vaccinia
virus DNA.

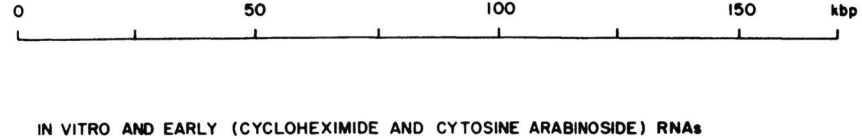

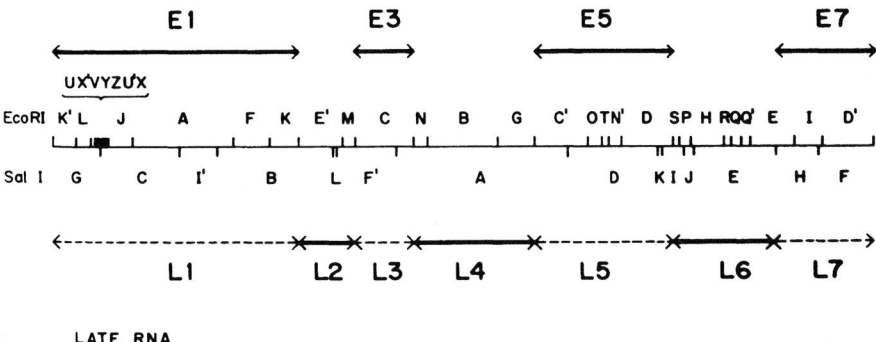

Fig. 3. Transcription map of ASF virus DNA. Reprinted with permission of (23).

VIRUS REPLICATION

    ASF virus infection starts with the interaction of the virus with a cell receptor. Titration experiments have shown the existence in both VERO cells and swine macrophages, but not in rabbit macrophages, of a receptor for the VERO cell-adapted virus. Virus penetration takes place by a mechanism of adsorptive endocytosis, that is sensitive to lysosomotropic drugs (A. Alcamí, A.L. Carrascosa, and E. Viñuela, unpublished data).

    Early RNA synthesized in ASF virus-infected cells in the presence of protein or DNA synthesis inhibitors hybridize preferentially to four regions in the genome, with coordinates E1 (0-51.9 kb), E3 (63.7-75.2 kb), E5 (100.1-111.6 kb) and E7 (150-170 kb) (Fig. 3). The RNA synthesized in vitro by the RNA polymerase associated with ASF virus particles hybridizes to the same DNA regions as early RNA. After hybridization selection with DNA restriction fragments and translation in reticulocyte lysates the RNA synthesized in vitro produces the same proteins as the RNA synthesized in infected cells in the presence of cycloheximide or arabinoside cytosine. This suggests that early RNA is synthesized in the infected cells by the virus-associated RNA polymerase (23).

After DNA replication, new RNA species arise that hybridize with DNA regions not transcribed in vitro or in infected cells in the presence of protein or DNA synthesis inhibitors (late RNA). These regions are L2 (51.6-63.7 kb), L4 (75.2-100.1) and L6 (111.6-150 kb). In addition, late RNA is complementary to regions L1, L3, L5 and L7 which also hybridize with early RNA, possibly due, at least in part, to the stability of some early RNAs (Fig. 3). A comparison of the translation products in vitro of early and late RNAs, selected with restriction fragments, has established which regions code for early transcripts only and for early and late RNAs. Many of the early messengers are present in an active form at late times after infection. This agrees with the observation that many early proteins are still synthesized after DNA replication. Genes for late proteins accumulate within a central region with coordinates 35.7-153.7, whereas the DNA ends code mainly for early proteins (23). Late RNA transcription is, like early RNA synthesis, host RNA polymerase-II independent, since the synthesis of both takes place in the presence of either α-amanitin or 5,6-dichloro-1, β, D-ribofuranosylbenzimidazole (J. Salas, M.L. Salas and E. Viñuela, unpublished data).

A two-dimensional gel electrophoresis analysis of lysates from ASF virus-infected cells, pulse-labeled with ^{35}S-methionine at different times of infection, has demonstrated 81 acid and 14 basic virus-induced polypeptides, whose molecular weight values range from 220 to 10 K. We could define three classes of ASF virus-induced proteins: one for which synthesis starts early after infection, continues for a period and then siwtches off; another for which the synthesis also starts early but continues for prolonged periods; and a third which requires DNA replication. Some of the virus-induced proteins undergo posttranslational modification, such as glycosylation (24,25, M. del Val and E. Viñuela, unpublished data), phosphorylation (25), and possibly protease-processing (18). The regulation pattern of the ASF virus-induced proteins is similar to that of poxviruses (26-28) and differs from that of frog virus 3, where the appearance of late proteins is not dependent on virus DNA replication (29).

Until now there are only two enzymatic activities that have been reported to increase in ASF virus-infected cells, a thymidine kinase (30) and a phosphonoacetic acid-sensitive DNA polymerase (31,32).

The replication of ASF DNA does not occur in enucleated cells (33). It is unclear if replicating viral DNA is present within the nucleus or only in the perinuclear region (34-36). An analysis of ASF virus replicating DNA molecules has shown the existence of head-to-head and tail-to-tail linked molecules (A. González, J.M. Almendral, A. Talavera and E. Viñuela, unpublished data) that may be replicative intermediates formed by some of the mechanisms proposed for poxvirus DNA replication (37,38).

TAXONOMY

Although the morphology (17), the lack of structural glycoproteins (15), the absence of neutralizing antibodies and the requirement of the cell nucleus for virus DNA replication (33) are properties that ASF virus and frog virus (29) have in common, many other properties of ASF virus are similar to those of poxviruses.

The DNA from both ASF and vaccinia virus is a unique sequence with hairpin loops and terminal inverted repeats (39-42). In contrast, iridovirus DNA is circularly permuted and has direct terminal repeats (43-45).

ASF and vaccinia virus particles contain the enzymes needed for the synthesis of virus messenger RNA (21,22,46,47). Moreover, RNA polymerase of either virion recognizes some, if not all, the early promoters present in either DNA (A. Talavera and E. Viñuela, unpublished data) and the RNA synthesis of early and late viral RNA is resistant to α-amanitin (J.Sa-

las, M.L. Salas and E. Viñuela, unpublished data). In contrast,
the presence of RNA polymerase in iridovirus is controversial
and the RNA synthesis in frog virus 3-infected cells is sensitive to α-amanitin (29). The messenger RNA from either ASF
or vaccinia virus-infected cells is polyadenylated. In contrast, the frog virus 3 specific messages are not polyadenylated. Finally, ASF and vaccinia virus induce the synthesis
of glycosylated proteins, whereas frog virus 3 does not (24,
48, M. del Val and E. Viñuela, unpublished data).

The ASF virus properties exclude it from all the families
defined by the International Committee for Taxonomy of Viruses
(49) and support the establishment of a new family, with ASF
virus as the only known representative.

GENETIC AND ANTIGENIC VARIATION

A possibility to account for the lack of detection of
neutralizing antibodies in sera from ASF virus-infected animals
is antigenic variation of critical antigens. Indications that
ASF virus undergoes antigenic changes are the following : 1)
pigs infected with attenuated viruses, derived from virulent
ones by passage in porcine leukocytes in vitro, are partially
resistant to the original but not to other virulent isolates
(50-53), 2) a soluble antigen, isolated from the spleen of
infected pigs, seems to be isolate-specific (54), 3) the hemad-
sorption-inhibition reaction (55-57), the complement-mediated
cytolysis (58), and the antibody-dependent cellular cytotoxy-
city (59), have made possible the distinction of different
virus isolates.

An approach to study virus variability is to compare
restriction patterns and maps of different virus isolates
(60,61). We have analyzed 23 ASF virus field isolates (9
African, 11 European and 3 American) by comparing the number
and size of their Sal I fragments. The comparison of the restriction maps derived for each isolate revealed the existence
of a central, highly conserved region that spans the majority
of the genome (∼125 kb), where few changes were detected (in
general, appearance and disappearance of restriction sites)

and two variable regions close to the DNA ends, which showed deletions or additions up to 8.6 kb (R. Blasco, M. Agüero and E. Viñuela, unpublished data).

Nucleotide sequence analysis of the variable regions has shown the existence of a multigenic family with homologous genes at either DNA end, close to the terminal inverted repeats (A. González, V. Calvo and E. Viñuela, unpublished data). A second gene family has been found between 10 and 20 kb from the left DNA end, within the most variable region of ASF virus DNA, with a complex pattern of repeats that seems to account for the deletions or additions found in that region in different virus isolates (J.M. Almendral, R. Blasco, F. Almazán, M. Agüero and E. Viñuela, unpublished data). The possibility that the existence of gene families in ASF virus DNA, a property not described until now for any other virus, is related to the virus escape from the host immune system will have to wait until the function of the putative proteins encoded within those regions is known.

The availability of a collection of monoclonal antibodies which recognize ten ASF virus proteins has allowed us to (1) determine the antigenic changes of ASF virus passaged in either porcine macrophages or monkey kidney (VERO) cells, (2) show the existence in a single pig of variants with different antigenic properties and (3) compare the binding properties of 23 field virus isolates (13).

Some proteins, like p72, are stable after passaging the virus in macrophages, whereas other proteins change after 20 (p17, p220/p150) or 10 passages (p27) (Fig. 4). The virus passaged in VERO cells changes much less than in macrophages.

From seven clones isolated from the spleen of an infected pig, one showed changes in protein p150, p37 and p14, other slight changes in protein p220/p150 and a third clone in protein p27 (Fig. 5).

In field virus isolates, we have detected antigenic changes in 6 out of 10 ASF virus proteins for which there are available monoclonal antibodies, suggesting a broad distribution of the variability (Fig. 6). The most variable proteins

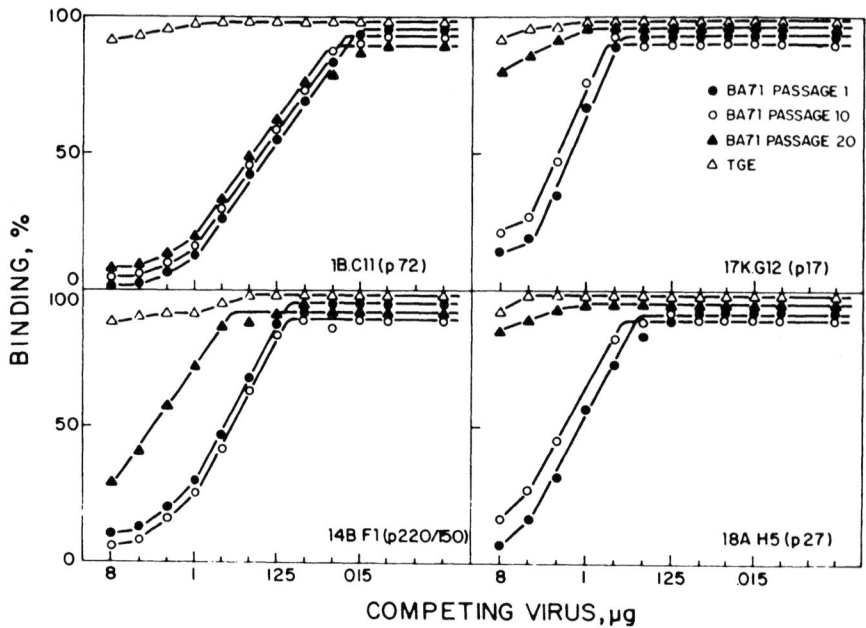

Fig. 4. Binding of monoclonal antibodies to ASF virus isolates passaged in porcine macrophages, texted by a competitive radioimmunoassay. The reference virus was the original virus clone of each isolate. FGE, transmisible gastroenteritis virus (TGE), an unrelated virus, was used as a control. Reprinted with permission of 13.

in the African isolates were p150 and p12. Protein p150 is a structural protein localized in the nucleoid in the virus particle and is antigenically related to protein p220 (18,19), a virus-induced, nonstructural protein that seems to be inserted in the membrane of infected cells (J.F. Santarén and E. Viñuela, submitted to publication). In contrast with the African isolates, protein p12 from the non-African viruses did not change. Protein p12 is a major structural protein, localized near the virion periphery (18,19).

The antigenic stability of protein p72 is important because it is the main antigen used in the detection of ASF virus-specific antisera by an enzyme immunoassay for the diagnosis of infected pigs (62, C. Vela, L. Enjuanes, A. Sanz

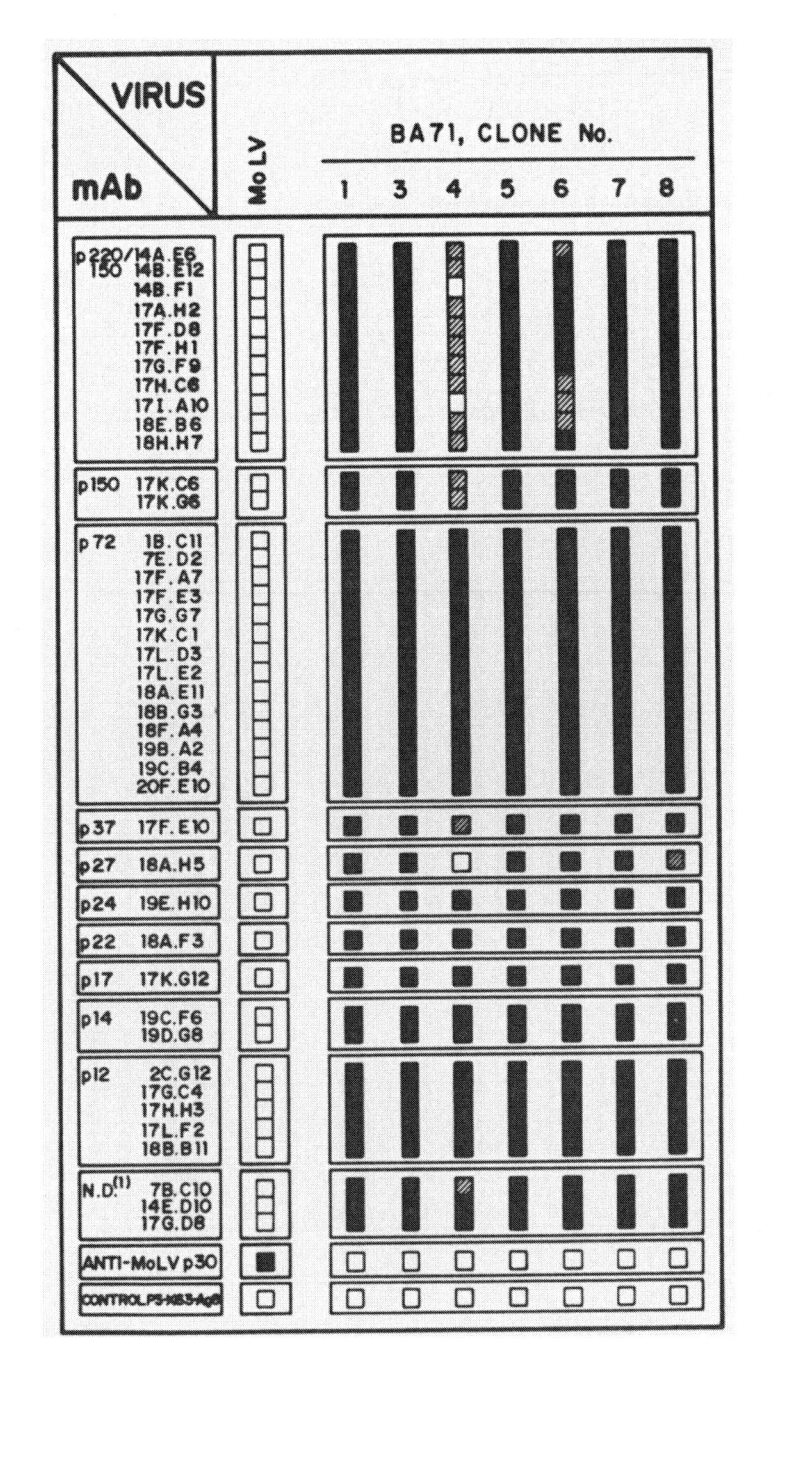

46

Fig. 5. Binding of monoclonal antibodies to ASF virus clones isolated from the spleen of a pig infected in the field with ASF virus BA71. The binding percentage value of each monoclonal antibody to each virus isolate, relative to the binding to a reference virus (BA71), was calculated by using the formula $100\ (C_{ij}-C_{1j})/(C_{ij}-C_{1j})$, where C_{ij} = cpm of monoclonal antibody i bound to the clone j (1,2,...7); C_{1j} = cpm of a blank monoclonal antibody specific for protein p30 from Moloney leukemia virus; C_{ij} = cpm of control of monoclonal antibody i bound to the reference virus 1; C_{11} = cpm of blank monoclonal antibody 1 to the reference virus 1.□ , 0-19; ▨ 20-40; ■ 41-100 binding percentage values, relative to the clone 1 binding value taken as 100. Reprinted with permission of (13).

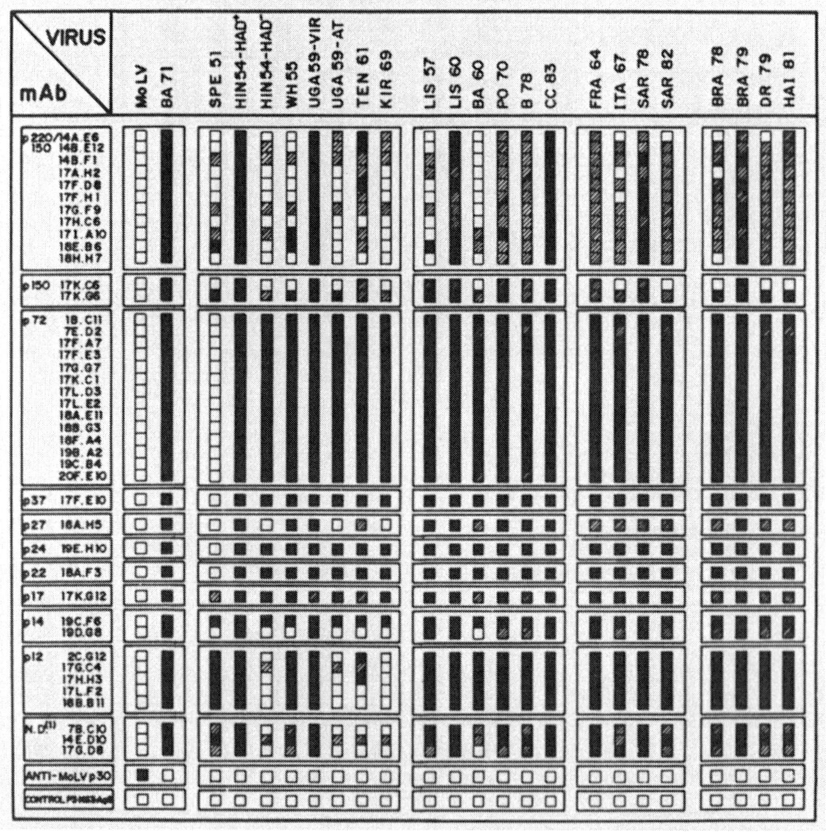

Fig. 6. Binding of monoclonal antibodies to African, European and American ASF virus field isolates. Symbols as in Fig. 5. Reprinted with permission of (13).

and E. Viñuela, unpublished data).

REFERENCES

1. Hess, W.R. Adv. Vet. Sci. Comp. Med. 25: 349-356, 1981.
2. Wardley, R.C., Andrade, C.de M., Black, D.., Castro Portugal, F.L., Enjuanes, L., Wess, W.R., Mebus, C., Ordás,A.,
   Rutili, D., Sánchez Vizcaíno, J., Vigario, J.D., Wilkinson, P.J., Moura Nues, J.F. and Thomson, G. Arch. Virol.
   76: 73-90, 1983.
3. Viñuela, E. Curr. Top. Microbiol. Immunol. 116: 151-170,
   1985.
4. Montgomery, R.E. J. Comp. Pathol. 34, 159-191, 243-262,
   1921.
5. Hess, W.R. In: Virology Monographs, vol. 9 (Eds. S. Gard,
   C. Hallauer and K.F. Meyer), Springer, Beling Heidelber,
   New York, pp 1-33, 1971.
6. Colgrove, G., Haelterman, E.O. and Coggins, L. Am J. Vet.
   Res. 30: 1343-1359, 1969.
7. DeBoer, C.J. Arch. ges. Virus Forsch 20: 164-179, 1967.
8. Malmquist, W.A. and May, D. Am. J. Vet. Res. 21: 104-108,
   1960.
9. Casal, I., Enjuanes, L. and Viñuela, E. J. Virol. 52:37-
   46, 1985.
10. Wilkinson, P.J. and Wardley, R.C. Br. Vet. J. 134: 280-
    282, 1978.
11. Moura-Nunes, J.F. and Nunes-Petisca, J.L. Comm. Eur.
    Comm. EUR 8466 EN, pp. 132-142, 1983.
12. Enjuanes, L., Cubero, I. and Viñuela, E. J. Gen. Virol.
    34: 455-463, 1977.
13. García-Barreno, B., Sanz, A., Nogal, M.L., Viñuela, E.
    and Enjuanes, L. J. Virol. 58, in press.
14. Carrascosa, A.L., del Val, M., Santarén, J.F. and Viñuela,
    E. J. Virol. 54: 337-344, 1985.
15. Del Val, M., Carrascosa, J.L. and Viñuela, E. Virology,
    151, in press.
16. Breese, S.S.Jr., and DeBoer, C.J. Virology 28: 420-428,
    1966.
17. Carrascosa, J.L., Carazo, J.M., Carrascosa, A.L., García,
    N., Santisteban, N. and Viñuela, E. Virology 132: 160-
    172, 1984.
18. Sanz, A., García-Barreno, B., Nogal, M.L., Viñuela, E.
    and Enjuanes, L. J. Virol. 54: 199-206, 1985.
19. Carrascosa, J.L., González, P., Carrascosa, A.L., Barre-
    no, B.G., Enjuanes, L. and Viñuela, E. J. Virol. 58,
    in press.
20. Adlinger, H.K., Stone, S.S., Hess, W.R. and Bachrach, H.
    L. Virology 30: 750-752, 1966.
21. Kuznar, J., Salas, M.L., and Viñuela, E. Virology 101:
    169-175, 1980.
22. Salas, M.L., Kuznar, J. and Viñuela, E. Virology 113:
    484-491, 1981.
23. Salas, M., Campos, J., Almendral, J.M., Talavera, A. and
    Viñuela, E. Virology, in press.

48

24. Tabarés, E., Martínez, J., Ruiz Gonzalvo, F. and Sánchez Botija, C. Arch. Virol. 66: 119-132, 1983.
25. Tabarés, E., Martınez, J., Martín, E. and Escribano, J.M. Arch. Virol. 77: 167-180, 1983.
26. Moss, B. and Salzman, N.P. J. Virol. 2: 1016-1027, 1975.
27. Esteban, M. and Metz, D.H. J. Gen. Virol. 19: 201-216.
28. Pennington, T.H. J. Gen. Virol. 25: 433-444, 1974.
29. Goorha, R. and Granoff, A. In Comprehensive Virogy, vol. 14 (Eds. H. Fraenkel-Conrat and R.R. Wagner), Plenum, New York, pp 347-399.
30. Polatnick, J. and Hess, W. Amer. J. Vet. Res. 31: 1609-1613, 1970.
31. Polatnick, J. and Hess, W.R. Arch. Ges. Virusforsch. 38: 383-385.
32. Moreno, M.A., Carrascosa, A.L., Ortín, J. and Viñuela,E. J. Gen. Virol. 93: 253-258, 1978.
33. Ortín, J. and Viñuela, E. J. Virol. 21, 902-905, 1977.
34. Vigario, J.D., Relvas, M.E., Ferraz, F.P., Ribeiro, J.M. and Pereira, C.G. Virology 33: 173-175, 1967.
35. Tabarés, E. and Sánchez, Botija, C. Arch. Virol. 61: 49-59, 1979.
36. Pan, I.C., Shimizu, M. and Hess, W.R. Am. J. Vet. Res. 41: 1357-1367.
37. Moyer, R.W. and Graves, R.L. Cell 27: 391-401, 1981.
38. Baroudy, B.M., Venkatesan, S. and Moss, B. Cold Spring Harbor Symp. Quant. Biol. 47: 723-729, 1983.
39. Ortín, J., Enjuanes, L. and Viñuela, E. J. Virol. 31: 579-583, 1979.
40. Sogo, J.M., Almendral, J.M., Talavera, A. and Viñuela,E. Virology 133: 271-275, 1984.
41. Garón, E.F., Barbosa, E. and Moss, B. Proc. Natl. Acad. Sci. US 75: 4863-4867, 1978.
42. Baroudy, B.M., Venkatesan, S. and Moss, B. Cell 28: 315-324, 1981.
43. Goorha, R., Murti, K.G. Proc. Natl. Acad. Sci. US 79: 248-252, 1982.
44. Darai, G., Anders, K., Koch, H.G., Delius, H., Gelder-blom, H., Samelecos, G. and Flügel, R.M. Virology 125: 466-479, 1983.
45. Delius, H., Darai, G. and Flügel, R.M. J. Virol. 49: 609-614, 1984.
46. Moss, B. In: Comprehensive Virol, vol. 3 (Eds. H. Fraen-kel-Conrat and R.R. Wagner), Plenum New York, pp 405-474.
47. Dales, A. and Pogo, B.G.T. In: Virology Monographs, vol. 18 (Eds. D.W. Kingsbury and H. Zur Hausen) Springer, Berlin, Heidelberg, pp 1-109, 1981.
48. Moss, B., Rosenblum, E.N. and Garon, C.F. Virology 46: 221-232.
49. Matthews, R.E.F. Classification and nomenclature of viruses, Karger, Basel, 1982.
50. DeTray, D.E. Adv. Vet. Sci. 8: 299-333, 1963.
51. Hess, W.R., Cox, B.F., Heuschele, W.P. and Stone, S.S. Amer. J. Vet. Res. 26: 141-146, 1965.
52. Pan, I.C., De Boer, C.J. and Heuschele, W.P. Proc.Soc. Expt. Biol. Med. 134: 367-371, 1970.

53. Stone, S.S. and Hess, W.R.   Amer. J. Vet. Res. 28: 475-481, 1967.
54. Stone, S.S. and Hess, W.R.   Virology 26: 622-629, 1965.
55. Coggins, L.   Prog. Med. Virol. 18: 48-63, 1974.
56. Vigario, J.D., Terrinha, A.M., Bastos, A.L., Moura Nunes, J.F., Marques, D. and Silva, J.F.   Arch. Ges. Virusforsch. 31: 387-389, 1970.
57. Vigario, J.D., Terrinha, A.M. and Moura Nunes, J.F.   Arch. Ges. Virusforsch. 45: 272-277, 1974.
58. Norley, S.G. and Wardley, R.C.   Immunology 46: 75-82, 1982.
59. Norley, S.G. and Wardley, R.C.   Res. Vet. Sci. 35: 75-79, 1983.
60. Wesley, R.D. and Tuthill, A.E.   Preventive Vet. Med. 2: 53-62, 1984.
61. Talavera, A., Almendral, J.M., Ley, V. and Viñuela, E. Comm. Eur. Comm. EUR 8466 EN, pp. 254-262, 1982.
62. Sánchez-Vizcaíno, J.M., Tabarés, E., Salvador, E. and Ordás, A.   Curr. Top. Vet. Med. Am. Sci. 22: 214-222, 1982.

6

CHARACTERIZATION OF ASFV PROTEINS

ENRIQUE TABARES

Departamento de Microbiologia, Facultad de Medicina, Universidad Autonoma de Madrid, Arzobispo Morcillo 4, Madrid 28029, Spain.

ABSTRACT

The analysis of polypeptides made in cells infected with African swine fever virus (ASFV) by high-resolution polyacrylamide gel electrophoresis revealed the synthesis of at least 44 polypeptides ranging in molecular weight from 9.5 K to 243 K. Most of the proteins are synthesized within the first eight hours after infection and can be classified as immediate-early, early or late proteins. Some are modified by the incorporation of prosthetic groups: of these, at least eight were phosphorylated, and three specific viral glycoproteins were detected by immunoprecipitation. The polypeptides are localized in the nucleus and cytoplasm of infected cells, but their function is unknown. They were classified as structural and non-structural polypeptides on the basis of their mobility in relation to the virion. Of the 28 polypeptides identified in intracellular virus, 23 are structural polypeptides. The other five could be cell polypeptides associated with the virus,as is the case with VP42 identified as cell actin. At least 34 polypeptides have been identified in extracellular virus where most of the major structural polypeptides correspond to those of the intra-cellular virus. The localization of these polypeptides in the virus was achieved by treatment with detergents and NaCl. No polypeptide that induces neutralizing antibodies has been found, but the major structural protein (VP73) could be important as an effective antigen in ELISA studies in the control of the disease.

INTRODUCTION

African swine fever virus (ASFV) causes a highly contagious and generally fatal disease of pigs (1-4). This virus, which is an icosahedral cytoplasmic DNA virus, is classified with the Iridoviridae (5). A characteristic of this family has been the failure to obtain neutralizing antibodies (5). In the case of African swine fever, however, pigs infected

with this virus may recover and resist challenge with the same strain of
ASFV, but the mechanisms of recovery and resistance have not yet been defined (6-8); therefore, a satisfactory vaccine has not yet been developed
either. Further knowledge of viral proteins can contribute to studies as
to their role in the immunology and control of the disease. With regard to
the latter, the major structural and infectious protein (VP73) is being
used as a diagnostic reagent (9) in epidemiological studies.

POLYPEPTIDE COMPOSITION OF VIRION
Purification of virus

The analysis of structural proteins has been done mainly using two
procedures for virus purification. These are based on centrifugation in
sucrose (10-13) or percoll gradients (14). The two procedures are in agree-
ment regarding the number and composition of the major proteins, although
they were done with either intracellular (13) or extracellular virus. The
criteria used to judge purification of intracellular virus were (i) morpho-
logical: intact and purified virus preparations as seen by electron microscopy (12); (ii) biochemical: contamination with cell DNA and proteins (15)
was measured, and it was only possible to detect polypeptides as phosphoproteins (16) and glycoproteins (13) in the purified virus preparations;
(iii) structural: the treatment with detergent and NaCL eliminated the
external envelope and the major structural protein VP73 (13). This treatment quantitatively conserves minor proteins with a molecular weight above
that of VP172, not described in extracellular virus (14). Although the
recovery of infectivity was poor, possibly because of the conformational
changes on the surface of the virus or poor virus recovery, the viral
particles were intact, and reproducible preparations were obtained. The
criteria for purification of extracellular virus were based fundamentally
on contamination with the cell vesicles and the infectivity of virion. The
first criterion was not applied to intracellular virus, although the treatment with detergents (NP40 or deoxycholate) plus a high concentration of
salt (0.5M NaCl) may dissolve the vesicles. The preparations in percoll do
not seem to be able to be reproduced because protein p45 differed in two
preparations (14, 17). The preparations appeared to be disrupted, as seen
by electron microscopy (18). The integrity of particles is more important
in the study of virus components than infectivity. The two methods, however,
produce similar patterns of structural polypeptides, at least in the major
viral proteins.

Number of structural polypeptides in the virion

It is of interest to study both intra- and extracellular virus in order
to understand the role of external viral proteins in the immunological response. At least 28 polypeptides have been identified in intracellular virus
with molecular weights ranging from 11.5 K to 243 K (13). The virus was
purified by centrifugation in NaCl-sucrose gradients, and the proteins were
analyzed by polyacrylamide gel electrophoresis. The polypeptides were
divided into three groups according to their relative abundance. The first
group consists of VP172, VP73 (the major structural protein), and VP15, which
are most abundant. (VP refers to the viral structural proteins and the number
to the molecular weight x 10^{-3}). The second group contains VP42, VP36, VP32,
VP25.5, VP12 and VP11.5 which are present in moderate amounts, and a third
group of polypeptides which occurs in lower concentrations. Some structural
proteins may represent cell components like VP42, identified as cell actin
(13) and VP232, VP219, VP92 and VP32. It is difficult to establish a relationship between intracellular and extracellular virus because their proteins may differ (20) and the virus strains and cell systems are also
different. The patterns of structural proteins are similar at least in the
major components. VP172, VP162, VP146, VP73, VP42, VP37, VP34, VP32, VP25.5, VP24.5
VP15, VP12 and VP11.5 (13), respectively, were purified from extracellular
virus by percoll gradients, corresponding to P150, p135, p130, p72, p45,
p37, p35, p34, p29, p27, p17, p14, and p10. The last-mentioned may overlap
with VP11.5 (13) or may, together with p125, form a component of extracellular virus. It is more difficult to correlate the minor components,
because they are not well defined in the extracellular virus protein gels
(14), where 34 polypeptides, ranging in molecular weight from 10 K to 150 K
were detected. In other studies of extracellular virus, purified in sucrose
gradients, the VP172, VP73, VP42, VP36 and VP32 of intracellular virus were
found to correspond with VP1 (125 K), VP2 (76 K), VP3 (50K), VP4 (44 K),
and VP5 (39 K) (10), and the components 1,6 (75 K), 9 (43 K), 10 (41 k) and
11 (33 K) (11). The differences in molecular weight found in various
laboratories were mainly due to the standard proteins used as a reference.
In some studies, proteins with molecular weights larger than 68 K (10, 11)
were not included, or the molecular weights of standard proteins such as
myosin were taken to be either 220 K (13) or 200 K (14). According to this,
the changes in molecular weight of IP243 and IP172 (13) are p220 and p150 (14).

Localization of polypeptides in the virion

ASFV is an enveloped virus with icosahedral symmetry (3, 18). The localization of polypeptides in the virion has been studied by treatment with detergents (10, 13) and salt (13). By treatment of intracellular virus with Nonidet P40 (NP40) and 2-mercaptoethanol (2ME), it was possible to obtain subviral particles (core I) that have lost some proteins and have a density in CsCl of 1.31 g/cm^3 which is higher than that of the complete virus (1.23 g/cm^3) (13). Under these conditions, the external envelope of the virion was removed, as seen by electron microscopy (13, 18). In some studies, particles with spikes or fiber-like projections were observed (15), but not in other studies on iridoviruses (18). The major components of this envelope seem to be the VP15 and VP11.5 (13) associated with another eight minor components. When treated with NP40, 2 ME and NaCl, the virus loses its major structural protein (VP73), and it is possible to obtain other subviral particles (core II) (13). This indicates that VP73 is the major component of the capsid (13, 18). The major component of the internal core is protein VP172 which is associated with eleven additional proteins (13). After analogous treatment of extracellular virus with NP40 without 2ME, viral fractions with polypeptides different in composition were obtained (10), and the viral particles did not have viral DNA (13). By reaction with specific monoclonal antibodies, VP73 has been localized on the surface of the virion (19). Studies on solubilization of viral proteins are important for understanding the immunological and enzymatic properties of viral components (20).

Modified structural proteins

The presence of glycoproteins was studied by the incorporation of ^{3}H-glucosamine and ^{3}H-fucose in the structural components. These two substances occurred in intracellular virus in three components that corresponded, in mobility, to the minor polypeptides VP89, VP56 and VP51 (13). VP51 had the highest content of the precursors. The specificity of these proteins, as viral proteins, can be confirmed by immunoprecipitation from infected cells (16). The three components are solubilized from the intracellular virus by treatment with NP40 and NaCl, suggesting their localization in the envelope or external surface of the virus. In extracellular virus, two regions of glycoproteins have been determined (11), but they seem to be unrelated to intracellular virus (13). Although the first group presents a similar electrophoretic mobility, the differences may be due to the electrophoretic system used.

The major structural protein in the envelope (VP15) can be classified as a phosphoprotein by the incorporation of ^{32}P-orthophosphate into the intra-cellular virus (16). Other possible structural phosphoproteins may be the polypeptides VP80, VP54, VP34 and VP24.5 (16).

Virion-associated enzymes

Several experiments (21, 22) have suggested that a DNA-dependent RNA polymerase associated with the virion exists, although no controls were reported in virus purification (21). 5% polyethelene glycol (PEG) was used to concentrate extracellular virus, and under these conditions the PEG could associate extraneous proteins with the virions (23). In addition RNA-modifying enzymes (22), topoisomerase (24), nucleoside triphosphate phosphohydrolases (22), and protein kinase (25) have been associated with the virion.

Immunological reactivity of structural viral proteins

In the experiments on fractionation of the virus by treatment with detergents, it was shown that the envelope proteins solubilized with NP40 and 2ME did not produce a reaction, or did so very poorly, with hyperimmune serum (13). This could be related to the failure to neutralize ASFV by polyclonal antibodies (26). The treatment of virus with NP40, 2ME and NaCl however, showed that at least proteins VP172, VP162, VP146 and VP73 act as antigens in the natural infection (13). Different treatments, with and without NaCl in the presence of NP40 and 2ME, have allowed preparations of about 90% purity to be obtained from the major structural protein (VP73) which is used as an effective antigen in the ELISA test (27). This test has been commercialized and can be very useful as a screening tool in the control of the disease. Antisera produced in pigs inoculated with VP73, however, do not have a neutralizing effect in vitro, and immunization does not protect pigs against the viral infection (28). Neutralizing monoclonal antibodies,with specificity against a cell protein,which seems to be a structural protein (p24) of extracellular virus, have been obtained (17).

SYNTHESIS OF VIRAL PROTEINS IN INFECTED CELLS

Effect of virus infection on cellular protein synthesis

The infection of MS cells cells with ASFV inhibits the synthesis of proteins that occurs from 4.5 hr pi (29). This suggests that the inhibition may be produced by a late protein and could be related to a structural component of the virus particle, as occurs in other DNA viruses (adenovirus,

frog virus 3, HSV-1 and vaccinia virus) (30). The shutoff is less rapid and
drastic than in poxvirus and herpesvirus, but it is clearly demonstrated
between 22 to 36 hr p.i. in relation to cell actin synthesis (16).

Identification and enumeration of viral proteins

ASFV DNA is a linear double-stranded (12, 31) cross-linked (32) molecule
with a molecular weight of about 10^8 (12, 30) which can code for the synthesis
of about 100 polypeptides of average size. Initial studies resolved 34 polypeptides in infected cells, ranging in molecular weight from 9.5 K to 243 K
(29). Infected cell polypeptides (IP) are defined by the following criteria:
a) stimulation in the rate of synthesis after infection, b) immunoprecipitation
by antisera against viral antigens, and c) differences in electrophoretic
mobility. (IP is expressed as the molecular weight of the polypeptide x 10^{-3}).
We have recently introduced a new emuneration - ICP (infected cell polypeptide)-
given in the order of polypeptides according to size (Table 1) (33). We have
numbered the IP in order to facilitate the comparison of proteins amongst
different laboratories, since different standard molecular weights apply to
the same protein, and in future studies by two-dimensional electrophoresis,
one band may correspond to several polypeptides: IP13 and IP12 produce at
least four spots of high intensity with a pK of about 6 (Urzaiqui et al.,
unpublished results). MS cells infected with ASFV and labelled with
^{14}C-amino acids were examined by high resolution sodium-dodecylsulphate,
polyacrylamide gel electrophoresis and showed 44 infected cell polypeptides
(16, 33). We have defined the IP as structural or non-structural polypeptides
on the basis of their electrophoretic mobility, as compared to polypeptides
from purified virus, pending a more complete characterization of these
proteins (Table 1).

Although the patterns of the IP are similar in different cell systems
(pig leukocytes, Vero, MS and BHK cells) infected with several virus strains
(33, 34, 35), differences in some polypeptides have been detected (36), such
as the change of IP10 to IP13.5 for the 608V55 strain (33, 34, 35). Changes
in the mobility of some proteins have been detected by immunoprecipitation (36).

Regulation of the synthesis of viral polypeptides

The synthesis of the different infected cell polypeptides is initiated
sequentially, and they are synthesized within the first eight hours after
infection (29). This agrees with the observation by electron microscopy of
viral structures in infected cells between 6 and 8 hr p.i. (12, 37, 38).
The IP can be classified as immediate early and early proteins,based on their

TABLE 1

ICP	IP	STRUCTURAL	GLYCO.	PHOSPHO.	I.EARLY	EARLY	LATE
1	243	+					+
2	172	+					+
3	162	+					+
4	146	+			+		
5	144				+		
6	122	+				+	
7	104				+		
8	97	+			+		
9	89	+	+				
10	80	+		+			+
11	78				+		
12	73	+					+
13	71						+
14	63	+				+	
15	56	+	+				+
16	54	+		+	+		
17	51	+	+			+	
18	48	+			+		
19	46	+			+		
20	39	+				+	
21	35	+(U)			+		
22	34	+		+	+		
23	31				+		
24	30			+	+		
25	27					+	
26	25.5	+			+		
27	25						+
28	24.5	+		+	+		
29	23.5				+		
30	23			+			+
31	21.5					+	
32	20.5				+		
33	19						+
34	18					+	
35	16.5					+	
36	16			+		+	
37	15	+		+			+
38	14.5						+
39	14				+		
40	13					+	
41	12	+					+
42	11.5	+					+
43	10						+
44	9.5				+		

(U) UNCERTAIN

requirement of protein synthesis for transcription (Table 1) (33), or as late proteins, based on their requirement for viral DNA synthesis (29, 33). The appearance of immediate—early proteins is sequential because only IP104, IP97, IP30, IP24.5 and IP9.5 were detected when the treatment with cycloheximide was done at a low multiplicity of infection (34), while at a high multiplicity, others, such as IP34 and IP23.5, were clearly detected in the same period (33). The major components of the internal core (VP172), capsid (VP73), and envelope (VP15 and VP11.5) are late proteins. This is in accordance with the lack of replication of the virus in cell cultures treated with phosphonoacetic acid (PAA) (39, 40).

Post-translational modification

Several types of post-translational modification of ASFV proteins, including glycosylation, phosphorylation, and possibly cleavage, have been reported. A comparison between the glycosylated polypeptides synthesized by infected and uninfected cells showed that there are at least seven polypeptides electrophoretically different from those of uninfected cells, with a molecular weight between 50 K and 90 K (16). The viral specificity of glycopolypeptides was tested by immunoprecipitation with specific antibody produced during the natural course of infection in swine. Hyperimmune pig serum specifically precipitated three polypeptides (gA, gB, gC) that have been related, by electrophoretic mobility, to the structural polypeptides VP51, VP56, and VP89, respectively (Table 1) (13, 16).

Eight phosphorylated IP have been detected in ASFV-infected cells (Table 1) and five were specifically immunoprecipitated with hyperimmune pig serum (IP80, IP54, IP34, IP24.5 and IP15). By their mobility, IP80, IP54, IP34, IP24.5 and IP15 could correspond to the minor structural viral proteins VP80, VP54, VP34, VP24.5, and one of the major structural proteins, VP15, which is easily localized by labelling with ^{32}P, in the preparation of purified virus (16). The phosphorylation of VP80, VP24.5 and VP15, which are localized in the viral envelope, may be related to the protein kinase activity, also localized in the viral envelope (25).

The maturation process of some proteins by cleavage has been suggested recently (17). Protein IP243 may be a precursor of the IP172, because they present a common epitope, detected by monoclonal antibodies (17, 19).

Distribution of polypeptides in the infected cell and enzymatic activities

Information on the distribution of polypeptides in the infected cell could be important for characterizing a specific polypeptide and for knowing

its function in the macromolecular synthesis of virus. Viral DNA synthesis
has been detected in the nuclei of infected cells (41). This may be related
to the failure to produce any viral proteins or DNA in enucleated cells (42).
Nuclear changes after infection include clumping of chromatin and the appearance of fibrillar structures (43). However, pulse-chase experiments with
infected swine monocytes indicate that the site of viral DNA synthesis is
the cytoplasm (44), although only the site of virus assembly in cytoplasm
could be detected by autoradiography after concentration of labelled viral
DNA. At least 21 viral polypeptides were found, totally or partially, in the
nuclear fraction obtained by treatment of infected cells with NP40 (16). The
presence of some of these polypeptides like IP243 and IP172 was confirmed
in the nucleus by immunofluorescence with monoclonal antibodies (17). Nuclear
viral antigens have been detected by immunofluorescence in the first hours
p.i. (45). These viral nuclear proteins may be operative in the replication
of viral DNA. This replication requires a virus-induced DNA polymerase (44)
which is sensitive to PAA (39). This polymerase and the other enzymes
possibly involved in viral DNA synthesis like thymidine kinase (46) need
further investigation.

Immunological reactivity of viral proteins

By immunoprecipitation, it was possible to determine that at least 25
polypeptides induce antibodies in the natural infection (16, 35). Proteins
IP243, IP172, IP162, IP146, IP97, IP80, IP73, IP54, IP51(gA), IP35, IP34,
IP27, IP25.5, IP24.5, IP23.5, IP23, IP15, IP14.5 and IP11 reacted as denatured
antigens, whereas IP89(gC), IP56(gB), IP31, IP30 and IP12 only reacted if
the antigen was not denatured (16, 35). This immunoprecipitation was very
important in the characterization of IP as viral polypeptides (16). The
IP12 seems to be the strongest inducer of antibodies in the natural infection,
and it is the major antigen in the immunoelectrophoresis (IEOP). This polypeptide may be a minor component of the intracellular virus with the same
mobility as VP12. This protein and VP73 seem to be very stable, because
they produce a reaction with all sera from infected pigs tested (28). Polypeptides IP73 and IP12 represent the major viral components in the infection
(9, 16) and may be of importance in their use as antigens in the serological
test for diagnosis. Pigs immunized with the cytoplasmic fraction from
infected cells produce antibodies against IP243, IP172, IP146, IP73, IP34,
IP14.5 and IP12. These antibodies, however, were not neutralizing. Sera
from pigs that survived the infection neutralized the cytopathic effect

in pig leukocytes and immunoprecipitated IP97, IP27, IP25.5 and IP15, whereas sera with negative neutralization did not react (35). This result may be important, but elimination of antibodies from a positive serum by adsorption to protein A, however, does not cause any alteration in its index of seroneutralization (35). Other studies have shown that at least 37 polypeptides induce antibodies in the infection, and it was possible to find differences in the molecular weights of some immunoprecipitable proteins among different isolates of virus, and in an isolate adapted to Vero cells as well as in one not adapted to these cells (36).

REFERENCES
1. Hess, W.R. Adv Vet. Sci. Comp.Med. 25, 39–69, 1981.
2. Sanchez Botija, C. Rev. Sci. tech. Off. int Epiz. 1, 991–1029, 1982.
3. Wardley,R.C, Andrade, C.M., Black, D.N., Castro Portugal, F.L., Enjuanes, L., Hess, W.R., Mebus, C., Ordas, A., Rutili, D, Sanchez vizcaino, Y., Vijario, Y.D. and Wilkinso P.J. Arch Virol. 76, 73–90, 1983.
4. Viñuela, E. Curr. Top. Microbiol. Immunol. 116, 151–170, 1985.
5. Matthews, R.E.F. Intervirol. 12, 129–296,1979.
6. De Boer, C.J., Pan, I.C. and Hess, W.R. J.A.V.M.A. 160, 528–522, 1972.
7. Ruiz Gonzalvo, F.R., Carnero, M.E. and V. Brugel. Comm. Eur. Comm. EUR 8466 EN, 206–216, 1982.
8. Bommeli, W.,Kihm, U. and F. Ehrensperger. Coom. Eur. Comm. EUR 8466 EN, 217–223, 1983.
9. Tabares, E., Fernandez, M.,Salvador Temprano, E., Carnero, M.E. and Sanchez Botija, C. Arch. Virol. 70, 297–300, 1981.
10. Black, D.N. and Brown, F. Y. gen. Virol. 32, 509–518, 1976.
11. Vigario, J.D., Castro Portugal, F.L., Ferreira, C.A. and Festas, M.B. Comm. Europ. Comm. EUR 5904 En, 469–482 , 1977.
12. Sanchez Botija, C., Mc Auslan, B.R., Tabares, E., Wilkinson, P., Ordas, A., Friedman, A., Solana, A., Ferreira, C.,Ruiz-Gonsalvo, F., Dalsgaard,C., Marcotegui,M.A, Becker, Y. and Schlomai, Y. Comm. Europ. Communities. EUR 5626e , 1977.
13. Tabares, E., Marotegui, M.A., Fernando, M. and Sanchez Botija, C. Arch. Virol. 66, 107–117, 1980.
14. Carrascosa, A., del Val, M., Santaren, J.F. and E.Viñuela. J. Virol. 54, 337–344, 1985.
15. Tabares, E., Gonzalvo, F.R., Marcotegui, M.A. and Ordas, A. Comm. Europ. Comm. EUR 590 EN, 507–531, 1977.
16. Tabares, E., Martinez, J., Martin, E. and Escribano, J.M. Arch. Virol. 77, 167–180, 1983.
17. Sanz, A., Garcia Barreno, B., Nogal, M.L., Viñuela, E. and Enjuanes,L. J. Virol. 54, 199–206, 1985.
18. Carrascosa, J.L., Caraso, J.M., Carrascosa, A.L., Garcia, N., Santesteban, A. and E. Viñuela. Virol. 132, 160–172, 1984.
19. Whvard, T.C., Wool, S.H. and Letehworth, G. J.Virol. 142, 416–420, 1985.

20. Payne, L. J.Virol. 27, 28-37, 1978.
21. Kuznar, J., Salas, M.L. and Viñuela, E. Virol. 101, 169-175, 1980.
22. Salas, M.L., Kuznar, J. and Viñuela, E. Virol. 113, 484-491, 1981.
23. Polsan, A. Prep. Biochem. 4, 435-456, 1974.
24. Salas, M.L., Kuznar, J. and Viñuela, E. Arch. Virol.77, 77-80, 1983.
25. Polatnick, J., Pan, I.C. and Gravell, M. Arch. ges. Virusforsch 44, 156-159, 1974.
26. De Boer, C.J. Arch ges. Virusforsch 20, 164-179, 1967.
27. Tabares, E., Fernandez, M., Salvador-Temprano, E., Carnero, M.E. and Sanchez Botija, C. Arch Virol. 70, 297-300, 1981.
28. Tabares, E., Fernandez, M., Salvador Temprano, E., Carnero, M.E. and Sanchez Botija, C. Comm. Europ. Comm. EUR 8466 EN 224-234, 1983.
29. Tabares, E., Martinez, J., Ruiz Gonzalvo, F. and Sanchez Botija C. Arch. Virol. 66, 119-132,1980.
30. Bablanian, R. Comprehensive Virol. 19, 391-429, 1984.
31. Enjuanes, L., Carrascosa, A.L. and Viñuela, E. J. gen. Virol. 32, 479-492, 1976.
32. Ortin, J., Enjuanes, L. and Viñuela, E. J. Virol. 31, 579-583, 1979.
33. Escribano, J.M. and E. Tabares (Submitted to publication).
34. Escribano, J.M. and E. Tabares. Comm. Europ. Comm., 1984,in press.
35. Tabares, E., Ruiz Gonzalvo, F. and Carnero, M.E. Libro jubilar prof. C. Sanchez Botija. Ed. Fareso. Madrid 1983.
36. Letehworth, G.J. and Whyard, T.C. Arch Virol. 80, 265-274, 1984.
37. Moura Nunes, J.F., Vigario, J.D. and Terrhinha, A.M. Arch. Virol. 49, 59-66, 1975.
38. Breese, S.S. and De Boer, C.J. Virol. 28, 420-428, 1966.
39. Moreno, M.A., Carrascosa, A.L., Ortin, J. and Viñuela, E. J.gen. Virol. 93, 253-258, 1978.
40. Gil Fernandez, C., Paez, E., Vilas, P. and Garcia Gancedo. Chemotherapy 25, 162-169, 1979.
41. Tabares, E. and Sanchez Botija, C. Arch. Virol. 61,49-59, 1979.
42. Ortin, J. and Viñuela, E. J. Virol. 21, 902-905, 1977.
43. Moura Nunes, J.F., Vigario, J.D. and Terrinha, A.M. Arch. Virol. 49, 59-66, 1975.
44. Pan, I.C., Schimizu, M. and Hess, W.R Amer. J. Vet. Res: 41, 1357-1367, 1980.
45. Polatrick, J. and Hess, W.R. Arch. ges. Viresforch 38, 383-385,1972.
46. Polatrick, J. and Hess, W.R. Amer. J. Vet. Res. 31, 1609-1613, 1970.

7

AFRICAN SWINE FEVER DIAGNOSIS

J.M. SANCHEZ-VIZCAINO

Instituto Nacional de Investigaciones Agrarias, Department of Animal
Virology, Embajadores, 68, 28012, Madrid, Spain

ABSTRACT

Since no treatment or effective vaccine against ASF is available,
the disease control is based on a rapid laboratory diagnosis and the
enforcement of strict sanitary measures. For these reasons, ASF diagnostic
procedures have received considerable attention over the last few years.

In this paper, the different techniques most frequently used for virus
identification and antibody detection in ASF diagnosis have been described,
as well as a general procedure for ASF diagnosis.

INTRODUCTION

The clinical diagnosis of African Swine Fever (ASF) presents diffi-
culties due to the great similarity of symptoms and lesions with other
haemorrhagic diseases of pigs, especially in those countries where both ASF
and hog cholera exist. Moreover, the clinical diagnosis of ASF has become
more complicated in recent years with the appearance of new clinical forms
characterized by slow and insidious clinical symptoms, low mortality, and
the presence of carrier animals. For these reasons, the laboratory diagnosis
of ASF is essential.

Since no treatment or effective vaccine against ASF is available, disease control has been based on a rapid laboratory diagnosis and the enforcement of strict sanitary measures. It is for these reasons, mainly, that
this aspect of ASF has received considerable attention lately.

As in other virus diseases, laboratory diagnosis of ASF can be based
on the demonstration of infectious virus, viral antigens, or antibodies
produced against the virus. At present, a variety of laboratory techniques
is available for both the detection of virus and the demonstration of
specific antibodies.

This paper reviews the different diagnostic techniques used for ASF
identification and isolation and for viral antigen identification and

antibody detection. A general procedure for ASF diagnosis is also described.

VIRUS IDENTIFICATION AND ISOLATION

Several techniques have been adapted for the identification of ASF virus (Table I). Some, such as the agar double diffusion test (1), complement fixation (2), immunoperoxidase (3), radioimmunosorbent assay (4), enzyme-linked immunosorbent assay (5),and electron microscopy, are not practical for routine diagnosis. The techniques most frequently used at present are: the hemadsorption test (6), direct immunofluorescence (7) for virus identification, and especially inoculations of pigs for virus detection.

Hemadsorption Test (HAD)

HAD is the universal test for virus detection and for the identification of new outbreaks. It is the most sensitive and specific test for ASF virus identification. However, this test is very laborious, taking one or more days to yield results; furthermore, it requires animals free of infection and appropriate facilities.

HAD is based on the hemadsorption properties of the ASF virus (6). This phenomenon is due to the adsorptive properties of swine erythrocytes to the membrane of infected macrophages, forming a characteristic "rosette" or "morula" configuration around these cells. Finally, hemadsorption exerts a cytopathic effect, resulting in destruction of the infected cells. Even though HAD is the most sensitive test for virus identification, it has been observed that a small number of field strains show only cytopathic effect without hemadsorption (8). To identify such strains, it is necessary to use direct immunofluorescence on the sediments of these cell cultures.

On the other hand, in Spain it has been observed that some samples from subacute or chronic forms need one or two subinoculations on leukocyte cultures before the hemadsorption phenomenon occurs. This may be due to the presence of antibodies blocking the hemadsorption reaction, which would explain why the virus requires one or two passages before hemadsorption becomes evident.

Direct Immunofluorescence (DIF)

This technique is the first method used for viral antigen detection. DIF is based on the demonstration of viral antigen on impression smears or in cut tissue samples from spleen, lung, lymph nodes or kidney through conjugated immunoglobulins against ASF virus.

TABLE I

TECHNIQUES ADAPTED FOR ASF VIRUS OR ANTIGEN DETECTION

CHARACTERISTIC PARAMETERS	HA	DIF	ELISA	RIA	CF	IP	EM	DNA-P
SENSITIVITY	HIGH	MODERATE	HIGH	HIGH	LOW	MODERATE	LOW	HIGH
SPECIFICITY	HIGH	HIGH	HIGH	HIGH	LOW	HIGH	LOW	HIGH
REPRODUCTIBILITY	HIGH	LOW	HIGH	HIGH	LOW	MODERATE	LOW	HIGH
PROCEDURE	EASY	EASY	MODERATE	NOT EASY	NOT EASY	MODERATE	NOT EASY	MODERATE
INTERPRETATION OF RESULTS	CAN BE DIFFICULT	DIFFICULT	EASY	EASY	EASY	EASY	DIFFICULT	EASY
TRAINING REQUIRED	MODERATE	MODERATE	MODERATE	MUCH	MODERATE	MODERATE	MUCH	MODERATE
USE IN FIELD CONDITIONS	NO	NO	YES	NO	YES	NO	NO	NO
STABILITY OF REAGENTS	POOR	GOOD	GOOD	POOR	POOR	GOOD	-	POOR
AUTOMATIZATION	NO	NO	YES	YES	YES	NO	NO	NO
COST	MODERATE	LOW	LOW	HIGH	LOW	LOW	MODERATE	NO
TIME REQUIRED	MUCH	MODERATE	MODERATE	MODERATE	MODERATE	MODERATE	MUCH	HIGH
APLICATION	CONFIRMATION OUTBREAK VIRUS ISOLATION	FIRST TECHNIQUE TO USE	NOT PRACTICAL	NOT PRACTICAL	NOT PRACTICAL	NOT PRACTICAL	NOT PRACTICAL	UNDER EVALUATION

HA : HAEMADSORPTION; DIF : DIRECT INMUNOFLUORESCENCE;
ELISA : ENZYME LINKED IMMUNOSORBENT ASSAY;
RIA : RADIO IMMUNO ASSAY; C.F : COMPLEMENT FIXATION.
IP : IMMUNOPEROXIDASE; EM: ELECTRON MICROSCOPY.

DIF is a very rapid and economical technique with a very high sensitivity index for acute forms of the disease. For subacute and chronic forms, the DIF test registers a sensitivity index of only 40%. This decrease in sensitivity seems to be related to the formation of antigen-antibody complexes in the fluorescent anti-virus antibody.

At present, and as a routine diagnostic procedure, the direct and indirect immunofluorescence (DIF and IIF) are performed together. This combination enables scientists to detect 85 to 98% of ASF cases.

Swine inoculation

Pig inoculation is essentially recommended to confirm the first outbreak in an ASF-free country, using two groups of pigs: one unvaccinated group and one vaccinated against hog cholera. After inoculation of material from the sick animal, the animals should be studied daily with respect to temperature and blood samples collected for leukocyte culture and antibody determination.

Recently, a new method for ASF virus detection in field samples has been developed. This method is based on the use of the recombinant plasmid pRPEL2 as viral reagent in the DNA-DNA hybridization test (9).

ANTIBODY DETECTION

Antibody detection in ASF presents a very special situation for two main reasons: firstly, the lack of vaccine antibodies, and secondly, the long period of time in which ASF antibodies persist in the sick animals.

Antibody determination is recommended for the study of subacute and chronic forms, carrier animals and large-scale epidemiological work.

Several techniques have been adapted for ASF antibody detection (Table II). Radioimmunoassay (4) presents problems when used as a routine technique, due to the radioactive material. As to the complement fixation test (2), the precomplementary activity of pig serum complicates the results. Today the most common techniques used are: indirect immunofluorescence (10), immunoelectroosmophoresis (11), and enzyme-linked immunosorbent assay (12).

Indirect Immunofluorescence (IIF)

This test is a rapid technique with a high sensitivity and specificity index for ASF antibody detection from either sera (10) or tissue exudates (13).

The IIF test is based on the detection of specific ASF antibodies which bind to a monolayer of cell lines infected by an adapted ASF virus that is

TABLE II

TECHNIQUES ADAPTED FOR ASF ANTIBODY DETECTION

CHARACTERISTIC PARAMETERS	IIF	IEOP	ELISA	RIA	CF	SAFA
SENSITIVITY	HIGH	LOW	HIGH	HIGH	LOW	HIGH
SPECIFICITY	HIGH	LOW	HIGH	HIGH	LOW	HIGH
REPRODUCTIBILITY	HIGH	HIGH	HIGH	HIGH	HIGH	HIGH
PROCEDURE	EASY	VERY EASY	EASY	NOT EASY	NOT EASY	VERY EASY
INTERPRETATION OF RESULTS	EASY	EASY	EASY	EASY	MODERATE	EASY
TRAINING REQUIRED	MODERATE	LITTLE	MODERATE	MUCH	MODERATE	LITTLE
USE IN FIELD CONDITIONS	NO	YES	YES	NO	YES	YES
STABILITY OF REAGENTS	GOOD	GOOD	GOOD	POOR	GOOD	GOOD
AUTOMATIZATION	NO	MODERATE	YES	YES	YES	YES
TIME REQUIRED	MODERATE	LITTLE	MODERATE	MODERATE	MODERATE	LITTLE
APLICATION	REFERENCE TECHNIQUE FOR SAMPLES	ONLY WHEN ELISA CAN NOT BE DONE	LARGE SCALE SEROLOGICAL STUDIES	NOT PRACTICAL	NOT USEFUL	COMPLEMENTARY TO ELISA

IIF : INDIRECT IMMUNOFLUORESCENCE; IEOP: IMMUNOELECTROOSMOPHORESIS.

ELISA : ENZYME LINKED IMMUNOSORBENT ASSAY; RIA : RADIO IMMUNO ASSAY.

CF : COMPLEMENT FIXATION.

used as the antigen. The specific antibody which is bound to the antigen is accordingly detected by a rabbit anti-pig IgG labelled with fluorescein, or, for better results, by a purified protein A from *Staphylococcus aureus* conjugated with fluorescein and used as an anti-porcine immunoglobulin G reagent.

During the past few years, IIF has proved to be more sensitive than DIF for ASF antibody detection in the subacute and chronic cases. This is probably due to the increase of antibodies in these clinical forms, which create antibody-antigen complexes that hinder the detection of the viral antigen. However, combination of the DIF and IIF techniques can detect from 85 to 95% of the Spanish ASF cases. IIF is recommended as a reference technique for ASF antibody detection,as well as for small-scale serological studies of subacute, chronic and carrier animals.

Immunoelectroosmophoresis (IEOP)

This technique has been used for large-scale serological studies, due to its being a rapid, uncomplicated, and low-cost method (11).

IEOP is based on the observation of a precipitation line between the antigen and the antibody when they are placed in an electrical field.

The sensitivity and specificity index obtained with IEOP is lower than that obtained with IIF or ELISA (14). At present IEOP has been substituted by the indirect ELISA test. However, in cases where ELISA cannot be used, IEOP could be recommended for large-scale studies, but positive results must be confirmed by IIF.

Enzyme-Linked Immunosorbent Assay (ELISA)

The indirect ELISA test is at present the most useful method for ASF serological studies. After the ELISA test was first adapted for ASF antibody detection (5, 15, 16), its sensitivity and specificity were increased with the incorporation of the viral protein Vp73 as an antigen (14, 17) and protein A from *Staphylococcus aureus* as conjugate (18).

ELISA is based on the detection of ASF antibodies bound to the viral protein by the addition of a protein-A conjugate to an enzyme that produces a visible color reaction when reacted with the appropriate substrate. Such visible reactions can be quantified by a photocolorimeter or read by mere human sight.

In addition to its high sensitivity and specificity already mentioned, ELISA has provided a highly reproducible test which is relatively easy to perform and interpret. Furthermore, its cost per sample is relatively low, and it can be automatized. All of these properties make this technique ideal for the study of a large number of animals.

ASF antibody detection automatic immunofluorescence

Recently, a rapid and simple semi-automated fluorometric assay (SAFA) for ASF antibody detection and quantification was adapted and evaluated (19). The test uses a hydrophilic polymeric colloidal matrix (colloimmune) coated with the ASFV structural Vp73 antigen as the solid phase, and the specific antibody is detected by a fluorescein-protein A conjugate. The sensitivity and specificity index obtained with SAFA is comparable to that obtained with IIF and/or ELISA. At present this technique is being evaluated under field conditions.

GENERAL PROCEDURE FOR ASF DIAGNOSIS

The fight against ASF must be based on a rapid laboratory diagnosis and the enforcement of strict sanitary measures.

The first study to be done in the laboratory should be the detection of viral antigen by direct immunofluorescence, as well as antibody detection, from sera or tissue sample exudates, by indirect immunofluorescence. These procedures take only two hours, and the combination of the two methods makes possible detection of 95 to 98% of ASF cases. This method is practical when the number of cases is not too large; otherwise, IIF should be replaced by the ELISA test.

Once positive results have been obtained with DIF and IIF (or ELISA) tests, one must check the origin of the samples. If they come from a new outbreak or from an area previously considered ASF-free, a sample must be inoculated into pig leukocytes for hemadsorption studies and viral isolation. To confirm results, this, as well as animal inoculation, must be done in an ASF-free country. If the samples originated in an enzootic area, positive results obtained by means of DIF and IIF are considered sufficient to warrant measures to be taken. If the DIF or IIF results are negative, a suspension of the sample (spleen, lymph nodes or lung) must be inoculated for hemadsorption studies. This inoculated tissue culture must be read every 24 hours. Three different results can be observed:

1. Positive hemadsorption and cytopathic effect. This is a clear case of ASF.

2. Negative hemadsorption and positive cytopathic effect. In this case, a direct immunofluorescence test should be done on the

TABLE III

SUMMARY OF GENERAL PROCEDURE FOR ASF DIAGNOSIS

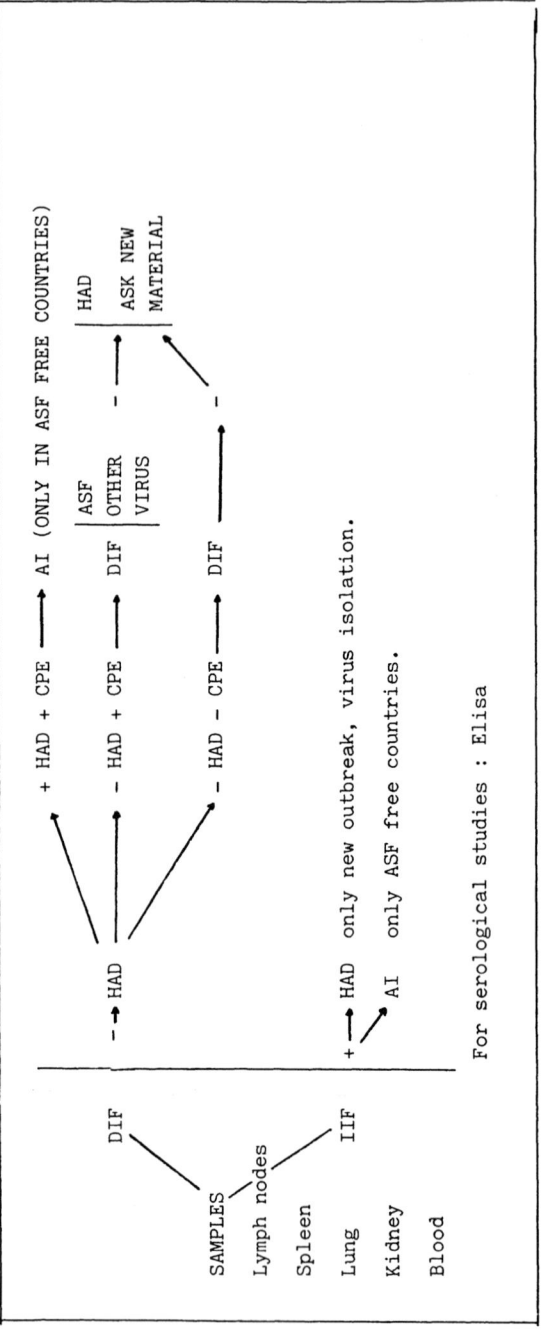

DIF : DIRECT IMMUNOFLUORESCENCE.

IIF : INDIRECT IMMUNOFLUORESCENCE.

ELISA : ENZYME LINKED IMMUNOSORBENT ASSAY.

HAD : HAEMADSORPTION.

CPE : CYTOPATHIC EFFECT.

cellular sediments to determine if it is a non-hemadsorping ASF strain or if the cytopathic effect is produced by another virus, such as Aujeszky.

3. Both hemadsorption and cytopathic effects are negative. In this case, if direct immunofluorescence is also negative, the culture should be subinoculated into a new leukocyte culture to detect the presence of ASF.

For serological diagnosis of carrier animals, surveillance of the disease, control of free areas and farms, slaughterhouse studies, and in general, large-scale antibody detection, the ELISA test is the recommended technique.

A summary of this general procedure is presented in Table III.

REFERENCES
1. Coggins, L. and Heuschele, W.P. Am. J. Vet. Res 27 (117): 485–488, 1966.
2. Cowan, K.M. J. of Immunology 86: 465–470, 1961.
3. Pan, I., Shimizu, M. and Hess, W. Am. J. Vet. Res. 39 (3): 491–497, 1978.
4. Wardley, R.C. and Wilkinson, P.J. Vet. Microbiol. 5: 169–176, 1980.
5. Wardley, R.C., Abu Elzein, E.M.E., Crowther, J.R. and Wilkinson, P.I. J. Hyg. Cam. 83: 363–369, 1979.
6. Malmquist, W.A. and Hay, D. Am. J. Vet. Res. 21: 104–108, 1960.
7. Bool, P.H., Ordás, A. and Sánchez Botija, C. Bull. Off. Int. Epiz. 72: 819–839, 1969.
8. Sánchez Botija, C. Rev. Sci. Tech. Off. Int. Epiz.: 1 (4), 1065–1094, 1982.
9. Caballero, R. and Tabarés, E. Spanish – American Seminar on ASFV. Madrid-Spain,1985.
10. Bool, P.H., Ordás, A. and Sánchez Botija, C. Rev. Patron. Biol. Anim. XIV (2): 113–132, 1970.
11. Pan, I.C., De Boer, C.J. and Hess, W.R. Cand. J. Comp. Med. 36: 309–316, 1972.
12. Sánchez-Vizcaíno, J.M., Crowther, J.R. and Wardley, R.C. CEE-EUR 8466: 297–325, 1982.
13. Sánchez Botija, C., Ordás, A. and González, J.G. Rev. Patron. Biol. Anim. 14: 159–180, 1970.
14. Sánchez-Vizcaíno, J.M. and Tabarés, E. CEE-EUR 8466: 101–105, 1982.
15. Hamdy, F.M. and Dardari, A.H. Vet. Record, 105: 445–446, 1979.
16. Sánchez-Vizcaíno, J.M., Martín Otero, A. and Ordás, A. Laboratorio, 440: 311–319, 1979.
17. Tabarés, E., Fernández, M., Salvador, E., Carnero, M.E. and Sánchez Botija, C. Arch. Virol. 70: 197–300, 1981.
18. Hortigüela, O. and Sánchez-Vizcaíno, J.M. Med. Vet. 5: 269–274, 1984.
19. Sánchez-Vizcaíno, J.M., Jacobson, R. and Arias, M.L. FAO/CEC consultation on ASF Research, Rome, Italy, 1984.

8

DIFFERENTIAL DIAGNOSIS BETWEEN AFRICAN SWINE FEVER AND HOG CHOLERA

C. Terpstra, DVM, PhD.

Department of Virology, Central Veterinary Institute,
P.O. Box 365, 8200 AJ Lelystad, the Netherlands.

Abstract

The highly varying picture of both African swine fever (ASF) and hog cholera (HC) makes a diagnosis of either disease on mere clinical and pathological grounds often impossible, particularly in subacute, chronic and atypical cases. Differences in clinical signs and post mortem lesions that might occur in acute and peracute outbreaks of the two diseases are listed. A tentative diagnosis of either ASF or HC should be confirmed by laboratory methods. A flow chart for differential diagnosis outlines the laboratory tests available and the timing at which results may be expected. A preliminary diagnosis of ASF in a hitherto free country should be based on the outcome of two different methods and be confirmed by an internationally recognized reference laboratory.

Introduction

The enzootic prevalence of African swine fever (ASF) on the Iberian peninsula since the late fifties, the incursions of the disease into France and Italy over the last two decades and its spread to Malta, Sardinia and some Latin-American countries in the late seventies have alarmed veterinary authorities of pig rearing countries all over the world. Most ASF-free countries have enforced veterinary legislative measures which prohibit imports of pigs and pig products from infected countries. Along these lines food and kitchen leftovers from aeroplanes, ships and international trains are destroyed. Control on the entrance of pig products, including food stuffs brought in by tourists, is nowhere absolute and the introduction of ASF into Belgium in 1985 has demonstrated that permanent vigilance should be exercised. In newly infected areas, an early diagnosis is of paramount importance and to a large extent determines the

Y. Becker (ed.), *African Swine Fever.* Copyright © 1987. Martinus Nijhoff Publishing, Boston. All rights reserved.

scenario and the prospects of an eradication programme. The initial outbreaks of ASF may easily be confused with classical swine fever or hog cholera (HC), especially in countries where the latter disease is enzootic. In Brasil and the Dominican Republic for example ASF spread under the cover of HC. A prompt diagnosis of ASF, therefore, implies its differentiation from HC. The occurrence of low virulent strains of both ASF and HC viruses, causing subacute, chronic, atypical and inapparent infections, makes it often impossible to diagnose or to differentiate the two diseases on the basis of clinical signs and pathological lesions. In cases of suspicion, laboratory investigations are indispensable in order to confirm or to rule out the possibility of ASF for certain.

The purpose of this chapter is to focus attention to the clinical and pathological differences of the two diseases that might be observed and to their differentiation by laboratory methods.

Clinical and pathological characteristics of ASF and HC

Although the symptoms and post-mortem lesions of ASF in general resemble those of HC, differences might be observed especially in acute and peracute fatal infections. Such outbreaks of ASF are characterized by high body temperatures while the animals behave normally for 1-3 days, followed by inappetance, a reddish and later cyanotic discolouration of the snout, ears, legs and/or abdomen, respiratory distress, vomiting and blood-tinged stools (1,2,3,4). The reddish discolouration of the skin and laboured breathing are less frequently observed in HC, whereas discharge of blood with the faeces is not recorded in HC (5,6). Signs of the central nervous system and coma are more common in the terminal stage of ASF than in HC. Haemostasis is severely impaired in ASF, resulting in secondary haemorrhages and protracted bleeding from intramuscular inoculation sites, and from the vulva in case of abortion (Biront, 1985, pers. comm.). Continued bleeding is rare in HC. Since the "carrier sow syndrome" is typical for infections with moderate or low virulent strains of hog cholera virus (HCV), its consequences in the form of the birth of piglets with congenital tremor, haemorrhages, malformation and/or oedema do not apply to ASF infections.

On post-mortem examination a variable amount of straw-coloured or reddish fluid may be present in the pericardium, and in the thoracic and abdominal cavities of pigs succumbed to ASF (1-4,7,8,9). Interlobular oedema of the lungs, oedema and congestion of the submucosa of the gall bladder and bile ducts are frequently observed in acute or subacute fatal cases of ASF. The same applies to the presence of blood clots in the gall bladder, bile ducts and renal pelvis. Unlike HC, the spleen is often partly or totally enlarged. In either disease a variable degree of haemorrhages, necrosis and reactive changes may be observed in the lymphatic tissues, notably in the gastrohepatic-, mesenteric- and renal lymph nodes. In advanced stages of ASF, however, these lymph nodes may resemble haematomas. Hyperaemia and petechial haemorrhages on the serous membranes, the epicard and endocard, the mucosae of the repiratory-, digestive- and urogenital tract are common in both diseases, but ecchymoses on these surfaces or subcutaneous and intramuscular haematomas are more in accordance with ASF than with HC.

The most significant difference between the histopathological lesions of the two diseases is the severe karyorrhexis of monocytes and macrophages in lymphoid tissues and lymphocytic infiltrates that occurs in ASF (7, 9). The massive destruction of lymphocytes resulting in an abundant residue of nuclear fragments is not observed in HC. A second, although less absolute, difference with HC is the degeneration of the epithelial cells of collecting tubuli in the kidney and the occlusion of these tubuli by amorphous proteinaceous casts in ASF.

For easy reference the clinical and pathological differences are listed in Table 1. It should be realised, however, that the clinical signs and pathological lesions of both diseases are highly variable and dependent on the virus strain involved. Consequently the differences listed are more quantitative than qualitative with the exception of tumor splenis and the haematoma – like visceral lymph nodes, which are characteristic for ASF. A tentative diagnosis of either ASF or HC should be based on the clinical observation and autopsy of several pigs and in any case has to be confirmed by laboratory methods.

Laboratory methods for diagnosis of ASF comprise the detection of virus by haemadsorption (10), of virus antigen by direct immunofluorescence (11) and detection of virus-specific antibodies in tissue fluid or serum. The prevalence of low virulent strains of ASF-virus (ASFV) in Spain has shifted the weight from detection of antigen towards antibody (12). For this purpose, either the indirect immunofluorescence test (ind. IFT) (13), immunoelectro-osmophoresis (IEOP) (14), ELISA (15) or the immunoperoxidase monolayer assay (IPMA) (Wensvoort, to be published) may be used. The ELISA and IPMA are the most sensitive. The latter test has the advantage that anti-ASFV-specific IgM or IgG can be detected as early as 5-7 and 7-10 days post-infection, respectively, by using conjugates of

Table 1. Clinical signs and post mortem findings in which ASF and HC might differ.

Feature	ASF (reference)	HC (reference)
Clinical:		
(sub-)cutaneous haemorrhage	common* (7,9)	rare* (5)
conjunctivitis	occasionally* (7) till common (4,9)	frequent* (5,7)
faeces with blood	occasionally (3,4,8,9)	not recorded (5,6)
haemostasis	severely impaired	reduced
involvement CNS**	common (4,8,9)	occasionally (5)
death in coma	frequent (8,9)	rare
congenital lesions	not applicable	tremor, oedema, malformations
Pathological:		
excess fluid in body cavities	frequent (1,2,7,8,9)	occasionally (5,8)
laryngeal haemorrhage	frequent (7,8)	common (5)
interlobular and subpleural lung oedema	common (1,7,9) till frequent (4)	rare (5)
splenomegaly	partly or totally (1,2,3,6,8,9)	absent (7,8)
haemorrhages visceral lymph nodes (notably gastro-hepatic)	frequent appearance of haematomas (4,8,9)	peripheral ("marbled")
epi(endo)cardial haemorrhages	ecchymosis common (1,4,7,8,9)	petechiae occasionally (5)
haemorrhages serous membranes	frequent and extensive (7)	rare (5)
gall bladder and bile ducts	usually distended; walls oedematous (1,2,4,7,9)	usually contracted; walls rarely congested (5,7)
renal pelvis with blood clots	occasionally (4,7,8)	rare (5,7)
cystic petechiae	common (7,9)	frequent (5)
Histopathology:		
karyorrhexis of monocytes	severe (7,9)	absent (7)
tubular degeneration with amorphous casts in renal medulla	frequent (7,9)	rare (7)

*	rare	< 5%	common	15-50%	** CNS = Central nervous system
	occasionally	5-15%	frequent	> 50%	

anti-swine isotype-specific monoclonal antibodies.

The laboratory procedures may somewhat depend on the disease situation in the country. The diagnostic and differential diagnostic procedures to be followed in case of a suspected outbreak of ASF or HC in a country free from one or both diseases are outlined in a flow chart (Fig. 1), which could serve as a guidance for laboratories responsible for diagnosis of exotic diseases. The procedure requires HC-immune pigs to be permanently available for differential diagnostic purposes. The same pigs may be used for preparation of leucocyte cultures, should the need arise. The laboratory should also stock direct fluorescent conjugates against ASFV and HCV, and fixed ASFV-infected cell cultures for ind-IFT and IPMA, antigen for IEOP, or an up to date ELISA kit. In countries where ASF is enzootic in the domestic pig population, investigation by one of the serological methods of contact-exposed pigs that have been sick for several days should have preferance above animal inoculation.

Samples of tonsils, spleen, lung, kidney, mesenteric and submandibular lymph nodes from dead or sacrificed pigs should be sent in leak proof packing to the laboratory by the quickest possible means. Cryostat sections of these organs are stained by direct-IFT with conjugates for HC and ASF (Fig. 1, tests 1,2). In case of a negative result, an appropriate dilution of tissue fluid is examined by indirect-IFT, or one of the other aforementioned tests suitable for detection of ASF antibodies (test 3). Meanwhile, a 20% suspension of tonsil and spleen in Hanks' balanced salt solution or Earle's minimum essential medium containing antibiotics and mycostatin at 10x the normal concentration, is kept for 1 h at room temperature and clarified by low speed centrifugation. The supernatant is mixed with a suspension containing approximately 2×10^6 PK-15 cells per ml and inoculated into Leighton tubes with coverslips for isolation of HCV (test 4). Cell sheets are fixed and examined by direct-IFT with HC-conjugate at 24 to 72 h post-seeding. HC-immune pigs are inoculated with supernatant fluid of the suspected tissue homogenate (test 5) the day after receipt of the specimens (day 1). One or more HC-susceptible pigs are inoculated as well and serve as controls. When an animal becomes febrile, at least 150 ml blood should be collected in heparinized tubes for the preparation of leucocyte cultures (test 6). The blood is pooled and the erythrocytes are left to settle down in a suitable cylinder held at 45° for 1.5 h at 37°C. The supernatant plasma and buffy coat are aspirated off, centrifuged for 10 min at 1500 rpm after which the leucocyte pellet is resuspended in approximately a third volume of plasma. After adding antibiotics and 100 μl of the homologous erythrocyte fraction, the suspension is dispensed into culture tubes

Fig. 1 DIFFERENTIAL DIAGNOSIS OF HC AND ASF
Method and (number of test)

Day

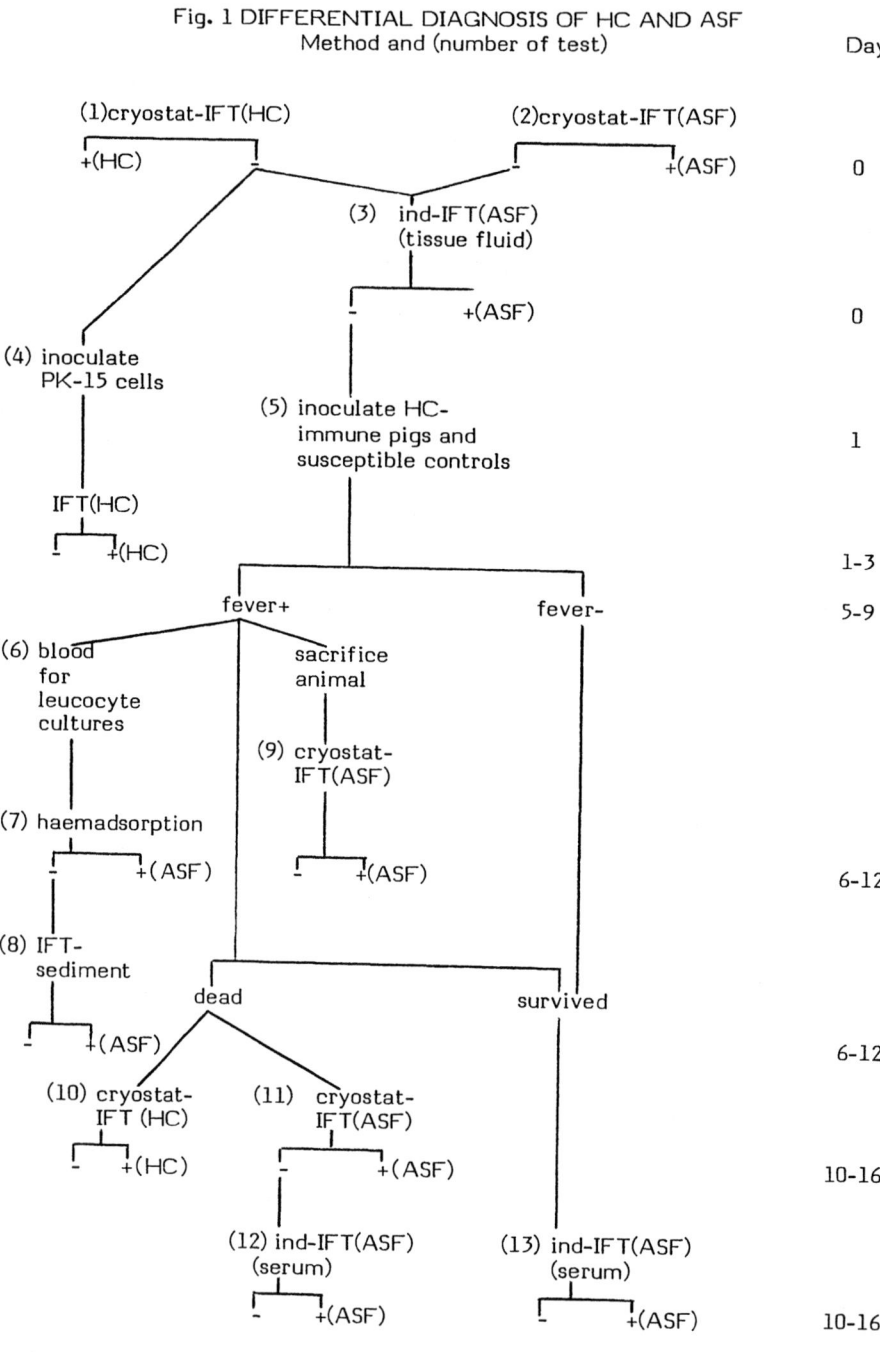

and incubated stationary at 37°C. The remaining plasma is stored frozen and used when the medium needs to be changed. Tubes are examined daily for haemadsorption (test 7). Low virulent strains of ASFV, however, may not show haemadsorption on primary isolation, as experienced in Spain, where 15-20% of the isolates appeared to have lost this property (16). If, therefore, no haemadsorption or cytopathic effect is observed within 72 h after inoculation, the leucocytes may be scraped from the glass, centrifuged and the sediment be examined for ASFV by direct-IFT (test 8). At least one animal with fever should be sacrificed and the organs be examined by direct-IFT for ASF (test 9), and in case of a control also for HC, while blood and tissue homogenates should be tested by haemadsorption (test 7). Pigs which die are examined for HC and ASF (tests 10, 11 and 12), similarly as field specimens, whereas serum of surviving animals should be tested for antibody against ASFV by the indirect-IFT or another suitable assay (test 13).

In view of the consequences for the national pig industry a preliminary diagnosis of ASF should be based on the outcome of two different methods. Generally, the responsible veterinary authorities will seek confirmation by an internationally recognized Reference Laboratory for ASF.

Whenever there is a strong suspicion of ASF in a hitherto free country and the diagnosis should take long, it is advisable not to await the outcome of the laboratory investigations, but to destroy all pigs on the suspected holding(s) and to take adequate veterinary police and zoo-sanitary measures to prevent any further spread of the disease.

References

1. Montgomery, R.E. J. comp. Path. 34: 159-191 and 243-262, 1921.
2. De Kock, G., Robinson, E.M. and Keppel, J.J.G. Onderstepoort J. vet. Sci. Anim. Ind. 14: 31-93, 1940.
3. Steyn, D.G. (quoted by Maurer et al.) Rep. Direct. vet. Educ. and Res. Onderstepoort 13: 415-428, 1928.
4. Polo Jover, F. and Sanchez Botija, C. Bull. Off. intern. Epiz. 55: 148-175, 1961.
5. Dunne, H.W. In: "Diseases of Swine", 2nd. edition, Ed. H.W. Dunne. Iowa State Univ. Press, Ames, Iowa, pp. 140-186, 1964.
6. Fuchs, F. In: "Handbuch der Virusinfektionen bei Tieren". Ed. H. Röhrer. Gustav Fischer Verlag, Jena, Band III/1, pp. 15-250, 1968.
7. Maurer, F.D., Griesemer, R.A. and Jones, T.C. Amer. J. vet. Res. 19: 517-

539, 1958.

8. Neitz, W.O. In: "Emerging diseases of animals". FAO Agricultural Studies 61: 1-70, 1963.

9. Moulton, J. and Coggins, L. Cornell Vet. 58: 364-388, 1968.

10. Malmquist, W.A. and Hay, D. Am. J. vet. Res. 21: 104-108, 1960.

11. Bool, P.H., Ordas, A. and Sanchez Botija, C. Bull. Off. intern. Epiz. 72: 819-839, 1969.

12. Sanchez Botija, C., Ordas, A., Gonzalvo, F. and Solana, A. In "Hog cholera/Classical swine fever and African swine fever". Proc. CEC/FAO Research Seminar, Hannover, Ed. B. Liess. EUR 5904, pp. 642-652, 1977.

13. Sanchez Botija, C., Ordas, A. and Gonzalez, J.G. Bull. Off. intern. Epiz. 74: 397-417, 1970.

14. Pan, I.C., De Boer, C.J. and Hess, W.R. Canad. J. comp. Med. 36: 309-316, 1972.

15. Sanchez Vizcaino, J.M., Crowther, J.R. and Wardley, R.C. In: "African swine fever". Proc. CEC/FAO Research Seminar, Sardinia, Ed. P.J. Wilkinson. EUR 8466, pp. 297-325, 1981.

16. Ordas, A., Sanchez Botija, C., Bruyel, V. and Olias, J. In: "African swine fever". Proc. CEC/FAO. Research Seminar, Sardinia, Ed. P.J. Wilkinson. EUR 8466, pp. 7-11, 1981.

9

SPONTANEOUSLY SUSCEPTIBLE CELLS AND CELL CULTURE METHODOLOGIES FOR AFRICAN SWINE FEVER VIRUS

In-Chang Pan

U.S. Department of Agriculture, ARS, NAA, Plum Island Animal Disease Center, P.O. Box 848, Greenport, New York, U.S.A.

ABSTRACT

The results of immunofluorescent studies revealed that those cells which belong to the mononuclear phagocyte system of van Furth or reticuloendothelial cells in lymphoid tissues and organs were susceptible to African swine fever virus (ASFV) infection. Furthermore, we have observed for the first time that the reticular epithelial cells of thymus and mesangial cells in renal glomeruli were also susceptible to ASFV infection, while the lymphocytes, neutrophils, vascular endothelial cells, hepatic cells, and tunica media of blood vessels were not susceptible to ASFV infection in vivo. The plastic adhering mononuclear cells (MNC) were stimulated by macrophage colony—stimulating factor (MCSF), and MNC-forming colonies were fully susceptible to ASFV infection. Infected pigs which died from acute fatal course seem to acquire a disseminated intravascular coagulation (DIC). Methodologies for cell cultures, application of cultured cells for virus infection, production of viral antigens for serodiagnosis, and use of various cell lines for growing ASFV are described and discussed.

INTRODUCTION

Following the successful isolation of African swine fever virus (ASFV) in swine bone marrow (BM) and buffy coat (BC) cell culture and the discovery of hemadsorption (HAd) reaction in infected swine leukocytes cultures by Malmquist and Hay (1) in 1960, attempts have been made to "attenuate" or modify ASFV by passage in cultures of spontaneously susceptible cells or by adaptation and passage in a variety of kidney cell lines.

After years of experimenting with ASFV in cell cultures and in swine, we have made a number of interesting observations. Here we present some of our unpublished observations along with a review of some highlights of our published articles relating to cell culture methods for ASFV. Also interpretations of some of the findings derived from the use of these methods are presented.

MATERIALS AND METHODS

Swine

Yorkshire and Tamworth swine of both sexes from a closed herd, meeting the requirements imposed by Plum Island Animal Disease Center, were used. About 20 swine of from 150 to 250 kg body weight were kept as blood donors. One liter of blood was taken at each bleeding, but they were never bled more than once a month. Pigs weighing 15 to 60 kg were used as required for each experiment.

Swine Infected with ASFV

More than 500 pigs experimentally infected with field isolates of ASFV (Tengani, Haiti, DR-II, and Lisbon 60), either died or were killed when moribund with acute ASF. They were necropsied and tissues from various organs were taken for virus culture, paraffin sections for H&E stain, and for frozen sectioning. Pigs which were preexposed to one virus and died or survived the challenge with homologous virulent virus were also included in the studies.

Culture Media

For culturing blood mononuclear cells (MNC) and various cell lines, following culture media were used:

Medium A: F15 (Grand Island Biological Co., Grand Island, NY) supplemented with 30% autologous porcine serum (APS), 15% Medium E, and antibiotics*. Heat-inactivated fetal bovine serum (FBS) may be substituted for APS.

Medium B: F15 supplemented with 30% APS, 15% of mouse L-cell culture fluid** (LCF), and antibiotics.

Medium C: 100% APS.

Medium D: F15 supplemented with 10% newborn bovine calf serum, and antibiotics.

Medium E: TC 199 (K.C. Biologicals, Lenexa, KA) supplemented with 10% fetal bovine serum (FBS), and antibiotics.

---

*antibiotics: 50 μg/ml of gentamycin sulfate (Garamycin, Schering Co., Kenilworth, NJ) and 2.5 μg/ml of Fungizone (R.R. Squibb and Sons, Princeton, NJ).
**Supernatant of mouse L-cells cultured in Medium E at 37 C for 4 days.

Buffy Coat (BC) Cell Cultures

The hemadsorption (HAd) test is a valuable diagnostic tool for ASF and is effective in testing specimens taken at all stages of the disease. The method described here is currently in use in our laboratory and is a modification of the method of Hess and Detray (2). Blood is withdrawn from the anterior vena cava of a pig into a flask containing glass beads. The flask is shaken as the blood is collected until defibrination is complete. When bleedings are from more than one donor, each blood is processed separately. The blood is then poured through sterile gauze to remove the beads and clotted fibrin, and 50 mg/l of Gentamycin sulfate and 2.5 mg/l of Fungizone are added. The defibrinated blood is centrifuged at 800 xg for 30 minutes in a swinging bucket rotor at room temperature. The serum is transferred to a flask containing a magnetic stirring bar. A 20 ml syringe and 13 gauge cannula is used to gather the BC layer which lies above the sedimented erythrocytes. The tip of the cannula is allowed to touch, but not dip below the surface of the packed cells and is moved back and forth across the surface as gentle suction is applied. Often the leukocytes adhere loosely to one another, forming a somewhat contracted mantle that may come off as a single mass without moving the cannula. The leukocytes collected are well dispersed in serum in conical 40 ml or 12 ml centrifuge tubes by repeated aspiration with a 20 ml syringe and cannula. The leukocyte suspension is centrifuged at 800 xg for 30 minutes at 5 C. A sharp demarkation between BC and erythrocytes makes easier drawing of BC without much contamination with erythrocytes. The BC and accompanying erythrocytes are added to the serum and are kept in suspension by magnetic stirring. The number of leukocytes per ml of suspension may vary over a considerable range and still yield cultures which are satisfactory for virus detection. As a simple rule to follow, 150 ml BC suspension may be prepared from 1 l of defibrinated blood by the procedures described.

The cell suspension is distributed as follows: 1 ml/Leighton tube (5 cm^2/tube); 0.4 ml/well for 24-well plastic plates (2 cm^2/well); 0.2 ml/well for 48-well plate (1 cm^2/well); 0.1 ml/well for 96-well plastic plates (0.36 cm^2/well). Plastic plates are recommended where a CO_2 incubator is available, because of the uniformity in quality of plates used. The Leighton tubes should be stoppered with rubber-lined screw caps, gum or silicone rubber stoppers. The cultures are incubated at 37

C. The plastic plates are kept in a humidified, CO_2 (3%) incubator at
37 C. The cultures may be usable within 24 hours and generally usable
as long as cell morphology remains intact (may be up to 8 days).
However, only cultures between 1 and 4 days are used for routine diagnosis, because of their uniformity in susceptibility to ASFV infection.

Hemadsorption test

Collection of specimens. Although virus may be isolated from practically any organ and tissue in acute ASF, spleen, gastro-hepatic lymph
nodes and whole blood are the preferred tissues to sample. They should
be collected aseptically from sick or recently dead animals. If they
can not be delivered to the laboratory within a couple of hours, they
should be chilled in ice or frozen immediately. When this is not feasible,
a minimum of 4,000 units of penicillin and 3 mg of streptomycin per ml
should be added to blood; other tissues may be immersed in phosphate –
buffered saline (PBS), consisting of NaCl 0.1369 M, KCl 0.00268 M,
Na_2HPO_4 0.0081 M and KH_2PO_4 0.00147 M, pH 7.2, or physiological saline
or any tissue culture medium containing antibiotics as described. The
specimens should be transported to the laboratory as quickly as possible.

Preparation of inoculum. In the laboratory, the solid tissues are
dispersed by grinding or are finely minced with scissors in a small
amount of PBS or any kind of tissue culture fluid containing antibiotics.
A 10% suspension is usually prepared. They are incubated at 37 C for 30
minutes before being centrifuged at 800 xg for 20 minutes at 5 C.

Inoculation of BC cultures with prepared specimens. Cultures 1-to
4-day old are most suitable for demonstrating typical HAd within 24
hours if the specimens have a titer of $10^{4.0}HAd_{50}/ml$ or higher. However,
cultures may be inoculated the same day they are prepared; in this case,
cells showing hemadsorption are smaller than those seen with older
cultures. At least 4 cultures should be inoculated, each receiving 1.0
ml of the tissue suspension for a Leighton tube, 0.4 ml/well for 24-well
plate, 0.2 ml/well for 48-well plate, and 0.1 ml/well for 96-well plate,
respectively. With spleen, excellent results may be obtained by placing
2 or 3 small tissue fragments (about 1 mm^3 in size) directly on the cell
layer. Uninoculated cultures serve as controls.

Hemadsorption reaction. When examining the BC culture, the Leighton
tube or the plastic plate should be rocked gently but sufficiently to
dislodge any loosely adhering erythrocytes. The Leighton tube is then

inverted to allow the red cells to drain away from the adhering leukocytes. The tube is next placed on the microscope with the side bearing the adhering cells uppermost. It is examined under 100 or 200 X magnification. Cultures in plastic plates are examined with an inverted microscope.

Virus titration with BC cultures

Buffy coat cultures are frequently used for virus titration in the laboratory. A ten-fold dilution of a test sample is made with Medium D. To test for viremia, citrated (0.38%), heparinized, or defibrinated blood samples are sonicated, centrifuged at 800 xg for 30 minutes at 5 C, and the top 1 ml is serially diluted with Medium D. Eight tubes or wells are used for each dilution of the test material. Inoculated BC cultures are observed for 10 days for HAd and cytopathic effects (CPE). Cytospin (Cytospin, Shandon Southern Product Ltd. England) preparations are made from all tubes or wells in dilutions in which some tubes or wells have varying degrees of CPE, and all tubes or wells of higher dilutions showing no evidence of HAd or CPE are also examined. Cytospin preparations are air dried, fixed with methanol for 2 minutes at room temperature, and stained with fluroescinated ASF antibody for detecting ASF virus antigen. The viral titer is expressed as the reciprocal of the dilution that indicated the HA_{50}/ml or $TCID_{50}$/ml (for non-hemadsorbing virus) as calculated according to the method of Spearman and Karber (see ref. 3).

HAd in leukocyte cultures prepared from the blood of the infected animal

Ten ml of blood are withdrawn aseptically in a syringe containing 0.3 ml of a 1.0% solution of heparin, 1,000 units of penicillin and 1 mg of streptomycin and transferred immediately to a tube and quickly mixed. The blood is centrifuged at 800 xg for 30 minutes at 5 C. The BC and 2 ml of plasma are drawn and mixed, and 1 ml each is placed in 2 Leighton tubes or 0.4 ml each for 4-wells in 24-well plates. After 5 or 6 hours incubation at 37 C, the cultures are examined microscopically. If it is negative for HAd, the plasma and floating cells are decanted, washed with Medium D, and replaced with 1.0 and 0.4 ml of Media A, respectively. The culture is returned to the incubator and examined daily thereafter for at least 8 days. If HAd occurs, it has the same appearance as that seen in the regular test.

Stimulation of plastic adhering, blood mononuclear cells (MNC) by macrophage-stimulating factor (MSF)

The glass-adhering MNC are grown in 6-well plates or T25 flasks, respectively, with Media A, B, C, containing 7 to 12 μCi/ml of [^{3}H]thymidine (20 Ci/mmole) for various lengths of time. Triplicate samples were harvested for monitoring DNA synthesis as an indicator of cell division. In the parallel studies, the susceptibility of MSF-stimulated cells of varying ages were assessed by inoculating ASFV, and the respective triplicate samples were harvested. The virus cultures were titrated by HAd reaction in BC. The changes in morphology of glass-adhering cells were assessed by observing slide cultures on which MNC were cultured with Medium A, B and C, respectively. They were observed unstained, or fixed and stained with Giemsa stain.

Swine alveolar macrophage cultures

A large quantity of a virus isolate (DR-I) which had not been adapted to Vero or other cell lines was produced in monolayer cultures of swine alveolar macrophages. Because we had reasons to believe that the virus adapted to Vero cells was a mixed population of selected clones of the similar nature or mixed with variants, viral DNA was extracted from virus grown in swine alveolar macrophage cultures for studying genomic variations among field isolates (4). The alveolar macrophage culture was chosen because an enormous number of spontaneously susceptible cells may be readily obtained from a single animal.

Young specific pathogen-free pigs, weighing approximately 60 to 100 Kg, reared with conventional husbandry, were anesthetized by intravenous injection of sodium thiopentothal (Pentothal, Abbott Lab., N. Chicago, ILL) usually given in an ear vein as 5 ml of 25% solution in PBS. After exanguination, the thoracic cavity was opened, the lung and heart were released from the parietal wall of the thoracic cavities, and the trachea with larynx was freed from the neck. Only lungs free of pneumonic lesions were used. The heart and esophagus were freed from the lung, and the entire trachea and lungs were blotted with a sterile towel to remove blood stains. A funnel of 7.5 cm diameter was inserted into the trachea through the larynx, and the neck of the funnel was ligated to the trachea with a piece of string. The funnel and lungs were kept at the vertical position, and 200 ml of the culture Medium E, containing 5 x of antibiotics and 0.2% of heparin, chilled to 5 C, were introduced into the diaphragmatic lobes of the lung through the funnel. Both diaphragmatic lobes were massaged with both hands for about one minute before the lavage fluid was discharged through the funnel into a container. For a second wash, the lungs were tilted 45 degrees and 200 ml of fresh culture medium was introduced. The frontal part of the diaphragmatic and cardiac lobes were massaged, and the contents were discharged. For a third wash, the posterior parts of the diaphragmatic lobes were lifted 45 degrees with the apecal lobes at the bottom, and 100 ml of the culture medium was introduced. The distended apecal lobes were massaged and the content discharged. In all, 350 to 400 ml of lung lavage fluid were recovered. To remove contaminating mucous, the lavage fluid was filtered through two metal sieves (50 and 100 mesh/square inch). The filtered lavage fluid was centrifuged at 400 xg for five minutes at 5 C, and the cell pellet was washed twice with the culture medium consisting of F15 supplemented with 50% calf serum and 2 x of antibiotics. An average of $3 \times 10^{9.0}$ cells of which approximately 60% were plastic adhering was obtained from the lungs of a pig. Forty ml each of cell suspension were distributed to six Falcon culture flasks (150 cm^2). After overnight incubation at 37 C, unattached cells were removed, and the monolayer of alveolar macrophages was washed three times with the culture medium. The attached cells were in close proximity to each other but never formed a confluent monolayer.

Cell Line Cultures

The following lines grew well in F15 supplemented with 2% BFS and 8% new-born bovine serum with Gentamycin (50 mg/l) and Fungizone (2.5 mg/l), except mouse L-cells — cultured in F15 containing 10% FBS and antibiotics: Vero cells (African green monkey kidney)

MS cells (African green monkey kidney)

MVPK cells (Porcine kidney)

CV1 cells (African green monkey kidney)

PK15 cells (Porcine kidney)

ST cells (Swine testicular cell)

These cells are generally seeded with a cell concentration of 2×10^5/ml in the culture medium in various plastic or glass containers as follows: T25 flask: 7 ml/flask.   6-well plate: 3 ml/well.

T75 flask: 30 ml/flask.   24-well plate: 1 ml/well.

T150 flask: 60 ml/flask. 48-well plate: 0.5 ml/well.

Leighton tube: 1 ml/tube. 96-well plate: 0.1 ml/well.

When cultured at 37 C in a CO_2 incubator, a confluent monolayer is formed within 2 to 3 days. In general, cell cultures were used by day 5 of culturing.

Adaptation of ASFV to Vero Cells

The method of adaptation of virus to Vero cells was the same as previously described (4). Briefly, ten ml of 20% spleen suspension in defibrinated, autologous blood with a virus titer of $10^{8.0}$ HAd_{50} or higher were sonicated (6 cycles of 10 sec each at maximum output at 5 C), centrifuged (800 xg, 30 minutes, at 5 C), and the supernatant was used as the inoculum. A Falcon culture flask (25 cm^2) containing a confluent monolayer of Vero cells, 3 to 4 days old, was washed once with CM; and 5 ml of inoculum was placed on the cells. The inoculated cell culture was incubated for 3 to 4 hours at 37 C, with redistribution of the inoculum every 10 minutes. At the end of adsorption, the inoculum was removed, and the cell sheet was washed with Medium E once and overlayed with 7 ml of Medium E. The inoculated Vero cell culture was incubated at 33 C to minimize inactivation of new virus progeny which is apt to occur during prolonged incubation at 37 C or higher temperature (Pan, I.C., unpublished data) and was observed every day for signs of CPE. In general, CPE was not observed during the first 2 passages. Cells were scraped from the plastic wall at 5 days post-inoculation (DPI) and suspended in the culture fluid. The cell suspension was treated with six 10 sec periods of sonication at the maximum output. After centrifugation at 800 xg for 15 minutes at 5 C, the supernatant was transferred to a new Vero cell culture; CPE was observed by 3 DPI at the 3rd passage. The virus was propagated through 3 more passages in Vero cells to produce a stock virus.

The L'60BM89-NHV-BC4-VR1 virus* was produced by a single passage of L'60BM89-NHV-BC4 in Vero cells. Extensive CPE occurred within 3 days after inoculation.

---

*L'60BM89-NHV-BC4-VR1 virus: L'60 virus serially passaged 89 times through swine bone marrow cell cultures (L'60BM89); the non-hemadsorbing virus (NHV) isolated from a pig infected with L'60BM89 virus was freed from contaminating hemadsorbing virus (HAV) by 4 generations of passage at limiting dilution in BC culture (L'60BM89-NHV-BC4) and was further passaged once in Vero cell culture to yield L'60BM89-NHV-BC4-VR1.

Cloning of virus by plaque purification

The method for plaque purification was previously described (5).

Production of various antisera, conjugation of antibody to fluorescein isothiocyanate (FITC) or horseradish peroxidase (HRPO)

Following antisera were produced either in rabbits or sheep and were conjugated to FITC (5) or HRPO (6) as previously described: swine anti-ASFV; rabbit anti-swine IgG (RAIg); ovine anti-swine IgG (OAIg); rabbit anti-swine complement (RAC); and rabbit anti-swine fibrinogen (RAF). The conjugation procedure was as previously reported (6) with some modifications. In our experience, sera of high antibody titers from swine, rabbit, and sheep were successfully conjugated as described below. Briefly, a crude Ig solution was obtained by 1/3 saturation with ammonium sulfate or Ig purified by affinity chromatography, and the total protein was measured by the Biuret method or by spectrophotometry (6). The protein solution was dialysed against 0.1 M carbonate buffer solution at pH 9.0. For conjugation with fluorescein isothiocyanate (Sigma Chemical Company, St. Louis, MO), the protein-to-dye ratio by wt. was 150:1. After conjugation for 90 minutes at room temperature with magnetic stirring, the excess dye was removed by gel filtration (Sephadex G-25, Pharmacia Fine Chemicals, Piscataway, NJ), and the highly charged protein, a source of nonspecific IF, was removed by DEAE-cellulose column chromatography. The protein conjugate was eluted from DEAE-cellulose with 0.01 M phosphate buffer solution containing 0.14 M NaCl, pH 7.3. The fraction that came out with the buffered saline front was collected. The protein-to-dye ratio of the conjugate (7) was usually between 1:1.3 and 1:2.1 The conjugate was clarified by centrifugation and then filtered through a 0.02 μm filter (Millipore Filter Corporation, Bedford, MA). Sodium azide (0.025%) was added to FITC conjugate as preservative and were stored at -20 C. Specificities of conjugates were determined as described (5).

A slightly modified version of the method of Nakane and Kawaoi (8) was used to couple anti-ASFV antibody and OAIg with HRPO as previously described (6,9).

Preparations of frozen sections and staining of specimens

Frozen sections (4 μm) were prepared in a cryostat microtome and stained as described (5). Demonstration of viral antigens in cytospin

preparations (10) and indirect immunofluorescent (IIF) test (6) have
been described.

Antibody detection by indirect immunoperoxidase plaque staining (IIPS) test

The preparation of antigen plates and test procedures were
described elsewhere (6).

New method for preparing ASF virus antigen for IEOP and ELISA tests

In our experience, when IEOP antigen prepared by the conventional
method (11) was used in the ELISA test substantial numbers of false-
positive reactions occurred (6). Therefore, a new method of preparing
antigen was developed as described (6).

ELISA Test

We performed the ELISA test as described elsewhere (7).

Immunoelectroosmophoresis (IEOP), electrophoresis (EP) and reverse immunoelectrophoresis (IEP) tests

The newly improved IEOP antigen prepared from L'60 virus adapted to
Vero cells was used for the IEOP test (11,12). For EP, swine sera were
electrophoresed on cellulose acetate strips and stained with ponceau S
(13). For a reverse IEP, a serum sample was electrophoresed (14) and
two troughs were cut into agarose gel along the path of current. To one
trough, the rabbit antiserum against swine whole serum was applied, and
ASF soluble antigens prepared for IEOP test was applied to the other
trough.

Collection of leukocytes from viremic blood for demonstrating ASF antigens

The heparinized blood (10 ml) from pigs in acute phase of ASF was
centrifuged at 400 xg for 20 minutes at room temperature. The BC layer
was aspirated and mixed with 3 volumes of a 0.83% aqueous solution of
ammonium chloride to lyse the residual RBC. After the mixture stood at
room temperature for 5 minutes, it was centrifuged at 200 xg for 5
minutes. The leukocyte pellet was suspended and washed 3 times in
Medium D by centrifugation. The washed leukocytes were then suspended
in 1 ml of bovine serum, and a cytospin preparation was made for
staining with anti-ASF conjugate.

Separation of MNC from peripheral blood

A slight modification of the ficoll-diatrizoate (F-D) gradient
centrifugation technique was used to separate the MNC from the other
leukocytes in blood (15).

RESULTS

Spontaneously Susceptible Cells in Pigs

Mononuclear phagocyte system. The results of IF studies on major organs and tissues of ASFV-infected swine largely agreed with those previously reported by Colgrove et al (16), Boulanger et al (17) and Moulton and Coggins (18) with some differences and new findings. Generally speaking, those cells which belong to the mononuclear phagocyte system of van Furth (19) or reticuloendothelial cells in lymphoid tissues and organs were infected viz: histiocytes in loose connective tissue (Fig. 1); Kupffer cells in blood sinusoid of liver; macrophages in alveolar septa and lumina (Fig. 2); free and fixed macrophages (reticulo-endothelial cells) in the red pulp of spleen (Fig. 3); free and fixed macrophages in the cell-poor substance of lymph nodes; free and fixed macrophages in the inter-follicular and subepithelial connective tissue of tonsil (Fig. 4); free macrophages in intertubular connective tissues of kidney; megakaryocytes, fixed and free macrophages in bone marrow; tissue macrophages of other major organs, including gastrointestinal tracts, the central nervous system, thymus, and the heart.

Reticular epithelial cells (REC) of thymus. The majority of cases with infected REC were pigs which succumbed to the disease between 8 to

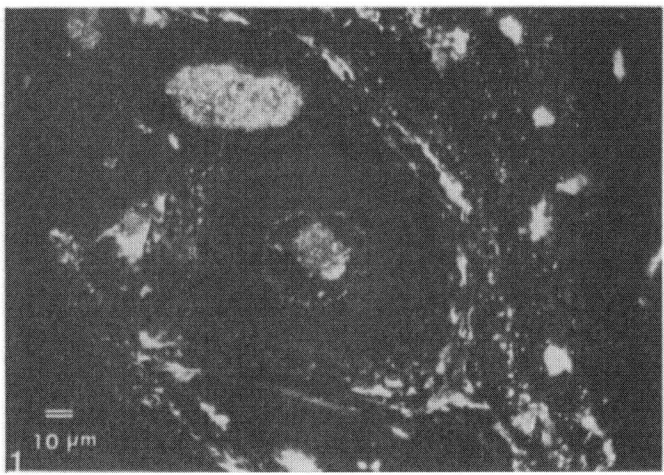

Fig. 1. Thrombi with ASFV antigens in muscular artery and vein. Elastica interna autofluoresced. Infected histiocytes were scattered in the loose connective tissue. Erupted portion of skin, frozen section, anti-ASFV conjugate stain. Case #9716. Died 7 DPI, 8th virus passage at limiting dilution of a vaccine virus (L'60BM89).

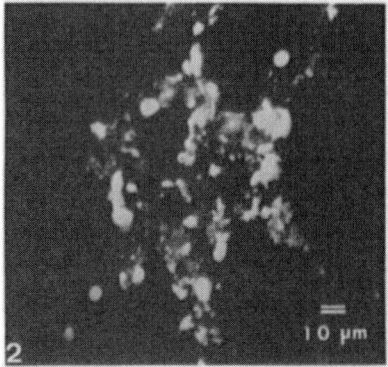

Fig. 2. Many macrophages in alveolar septa and few in alveolar sacs were infected with ASFV. Edematous lung, frozen section, anti-ASFV conjugate stain. Case #742. Died 15 DPI, Haiti-1 virus.

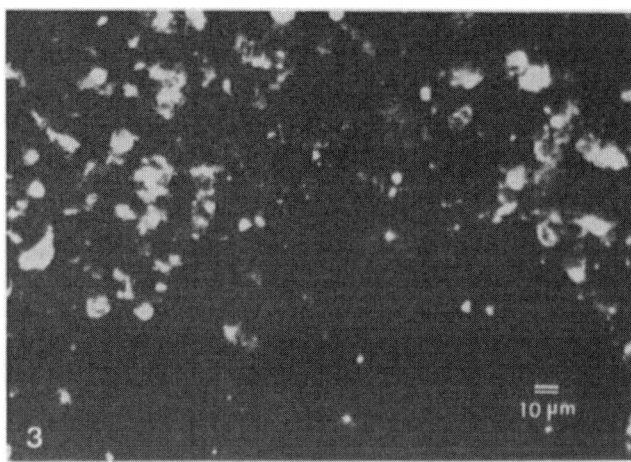

Fig. 3. Numerous macrophages infected with ASFV in red pulp of spleen. Notice white pulp at lower half of the field was essentially free from infected cells. Spleen, frozen section, anti-ASFV conjugate stain. Case #9716.

10 DPI with little difference in the virulence or dosage of the ASFV used. However, it was difficult to detect infection of REC in pigs that were exposed to virus isolates of lesser virulence, e.g. BR-1 or DR-2 (10), that died later than 12 DPI. Dose dependency of the clinical course in infections with moderately virulent viruses was reported previously (20). On the other hand, the Tengani virus isolate (10,20), which is the most virulent isolate in our collection, normally kills pigs within 5 to 8 days after administration of 100 HAD_{50} units of virus, but usually REC do not show signs of infection. In pigs inoculated with L'60 virus, REC infection was usually evident.

The infected REC showed strongly fluorescing clusters or islands of large cells with numerous cytoplasmic projections which were seen embracing many non-fluorescing lymphocytes in the immediate vicinity (Fig. 5). Infrequently, REC of Hassal's corpuscles also contained fluorescing ASFV antigens (Fig. 6).

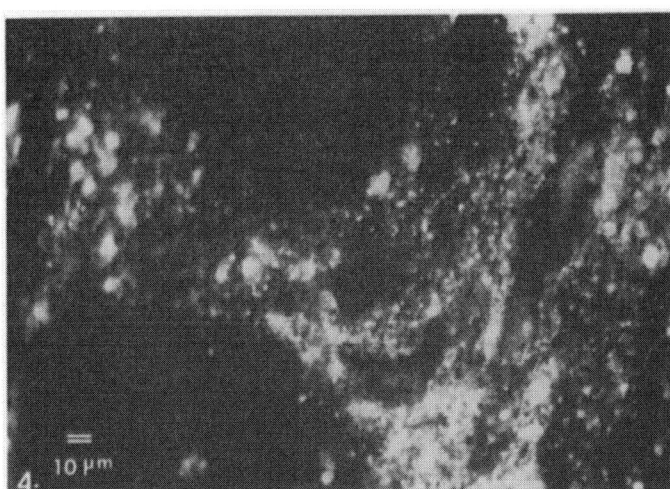

Fig. 4. Numerous macrophages were infected and disintegrated in interfollicular connective tissue in subepithelial layer of tonsil. Lymphocytes in B dependent area in two lymph follicles were not infected. Tonsil, frozen section, anti-ASFV conjugate stain. Case #167. Died 9 DPI, Tengani virus.

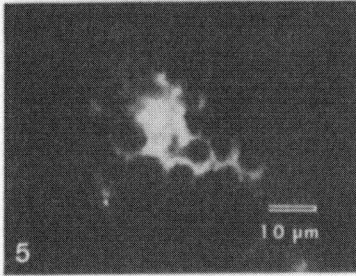

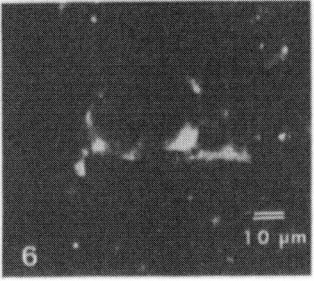

Fig. 5. Infected reticular epithelial cell of thymus with numerous cytoplasmic projections embracing non-fluorescing lymphocytes in the vicinity. None of lymphocytes (T cells) were infected. Thymus, frozen section, anti-ASFV conjugate stain. Case #111. Killed 14 DPI, Madrid '75 virus.

Fig. 6. Reticular epithelial cells of a small Hassal's corpuscle in medulla of thymus were infected. Thymus, frozen section, anti-ASFV conjugate stain. Case #8079. Killed 10 DPI, L'60B6-BC10 virus.

Mesangial cells of renal glomeruli. The glomerular mesangial cell, which is the fixed phagocyte in the mesangium of the glomerulus, was infected in the majority of pigs that died from acute ASF (Fig. 7a). Simultaneously, sandy or granular fluorescing ASF antigens (Fig. 7b) were seen dispersed throughout the mesangial area and in microthrombi of

94

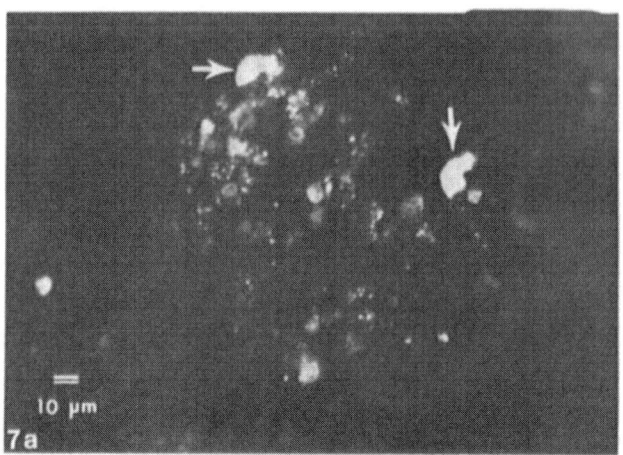

7a

10 μm

Fig. 7a. Few
mesangial cells
infected with
ASFV (arrow)
with relatively
intact
morphology
were seen in
the glomeru-
lar tuft.
ASFV antigens
of sandy or
granular
appearence
were seen in
glomerular
capillary
loops. Kidney,
frozen section,
anti-ASFV
conjugate
stain. Case
#502. Died 8
DPI, HT-NHV-
BC4 virus.

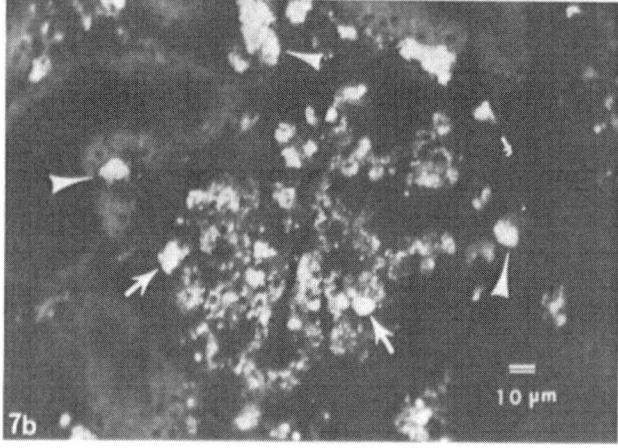

7b

10 μm

Fig. 7b.
Numerous clumps
of ASFV anti-
gens in a
form of
microthrombi
or emboli
(arrow) were
seen in
capillary
loops in the
glomerular
tuft as well
as in the
intertubular capillary beds (arrow head). Infected mesangial cells with
intact morphology were not seen. Kidney, frozen section, 4 uM thick,
methanol fixed, anti-ASFV conjugate stain. Case #168. Died 8 DPI,
Tengani virus.

capillary loops in the glomerular tuft (Fig. 7c) and in the intertubular

capillary beds (Fig. 8). The distribution of ASFV antigens in the

kidney of 29 pigs infected with four field isolates are shown (Table 1).

Blood vessels. Fluorescence for ASF antigens has been reported in

blood vessels in the terminal stage of the disease (16). In rare occa-

sions in our cases, blood vessels were found to contain abundant ASFV

antigens (Fig. 9), immunoglobulins (Fig. 10), and complement (Fig. 11),

Table 1. Distribution of ASFV Antigens in Kidney.

Pig No.	Virus	Dose HAD50	Route	Duration sickness	Glomerular Tuft mesangial cell/area	capillary loop	micro-thrombi	Intertubular space microthrombi in capillary beds	Histiocytes
1	Tengani	0.4	I.M.	4	-	-	+	+	+
2	"	4.2	I.M.	2	+	-	+	+	+
3	"	4.2	I.M.	2	-	-	+	+	+
4	"	4.2	I.M.	2	+	-	+	+	+
5	"	42.0	I.M.	3	+	+	+	-	+
6	L'60	316	I.M.	5	-	-	+	-	-
7	"	316	I.M.	6	+	-	+	-	-
8	"	1.2×10^{10}	I.V.	1	-	-	-	-	-
9	"	1.2×10^{10}	I.V.	2	+	+	+	-	+
10	"	1.2×10^{10}	I.V.	3	-	+	+	+	+
11	"	1.2×10^{10}	I.V.	3	+	+	+	+	+
12	"	1.2×10^{10}	I.V.	3	+	+	+	+	+
13	HT-NHV	5×10^{8}	I.V.	3	+	+	+	-	+
14	-BC4	NT	Contact	7	+	+	+	-	+
15	"	NT	Contact	4	+	+	+	-	+
16	"	NT	Contact	3	+	+	+	-	+
17	"	10	I.M.	7	+	+	+	-	+
18	"	100	I.M.	14	+	+	+	+	+
19	"	1,000	I.M.	7	+	+	+	-	+
20	"	10	I.M.	7	+	+	+	-	+
21	"	10	I.M.	5	+	+	+	-	+
22	"	2×10^{8}	I.V.	2	+	+	+	-	+
23	HT-I	3×10^{9}	I.V.	2	+	+	+	-	+
24	"	3×10^{9}	I.V.	2	+	+	+	-	+
25	"	3×10^{9}	I.M.	11	+	+	+	+	+
26	"	3	I.M.	13	-	+	-	-	-
27	"	32	I.M.	14	+	+	+	-	+
28	"	2×10^{9}	I.V.	4	+	+	+	-	-
29	"	3×10^{9}	I.V.	6	+	+	+	-	-

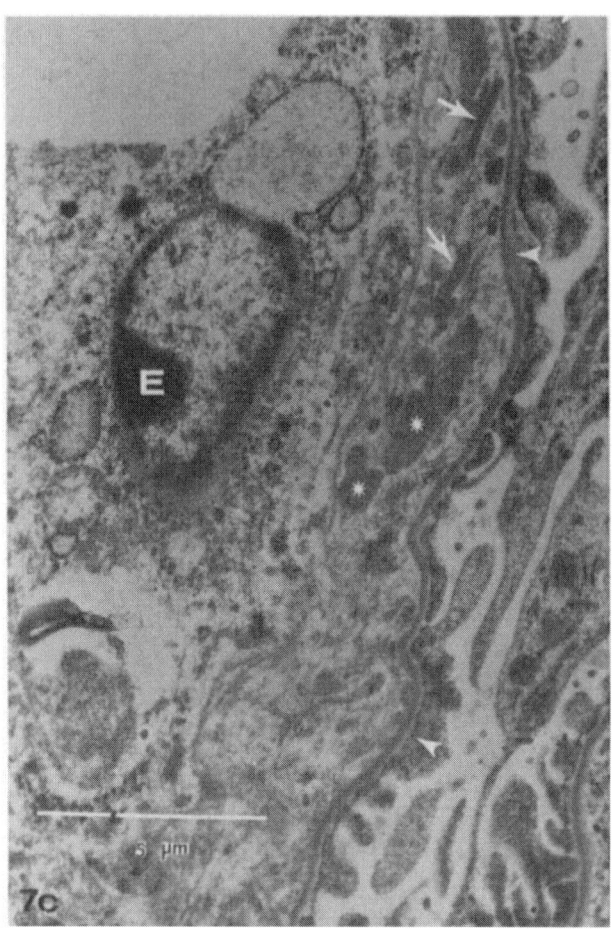

Fig. 7c. Subendothelial deposition of flocculent and electron dense deposits in glomerular capillary loop. The electron-dense deposits were presumably immunune complexes (*) and incompletely polymerised fibrin (arrow). Notice many cell debris in capillary lumen, degenerative endothelial cell (E), and compromised foot processes (arrow head) of glomerular epithelial cells.

respectively. In many cases, presence of fibrin thrombi was apparent (Fig. 10). There was no evidence that endothelial cells of various size blood vessels were infected. Occasionally seen, however, were fluorescing amorphous materials (presumably a viral antigen-antibody-C-platelet complex) adhering on the surface of endothelial cells of large vessels.

T and B lymphocytes in lymphoid organs. Lymphocytes were devoid of infection. To prove this point, four organs were selected for observations:

Thymus and spleen. The lymphocytes in thymus and in periarterial

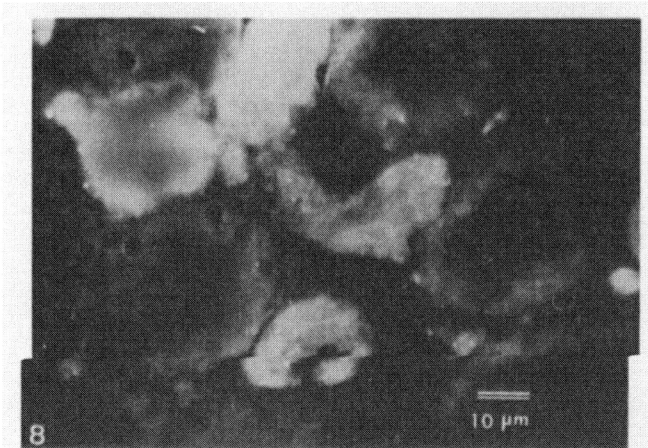

Fig. 8. Massive amount of ASFV antigens in fibrin thrombi occluding capillary lumens in the inter-tubular area. Tubular epithelial cells were autofluorescing. Kidney, frozen section, anti-ASFV conjugate stain. Case #168.

lymphatic sheath of spleen (T dependent area) were used as source of T lymphocytes. None of the lymphocytes in histologically relatively intact thymus (Fig. 5) and in periarterial sheath of spleen (Fig. 3) fluoresced for ASFV antigens; macrophages and reticularendothelial cells fluoresced strongly in the red pulp (Fig. 3).

Tonsil and lymph node. Lymphocytes in tonsilar lymph follicles (Fig. 4) and lymph follicles of regional lymph nodes (B dependent area) did not fluoresce for ASFV antigens, while macrophages and reticulum cells fluoresced strongly (Fig. 4).

Blood leukocytes. Based on the following observations, the claim that lymphocytes (16,21,22) and neutrophils (16,23) are susceptible to ASFV infection was not substantiated.

Cultured normal leukocytes. The two fractions of heparinized blood of normal pigs were obtained in F-D centriguation (15). More than 95% of the top fraction contained viable lymphocytes and monocytes. The cells were washed twice with Medium D by centrifugation at 400 xg for 10 minutes at room temperature. The bottom fraction contained granulocytes and red cells. After lysing red cells, granulocytes were spun down by centrifugation at 400 xg for 10 minutes. The cells collected were washed twice with Medium D. Then, cells from each fraction were cultured separately with autologous serum overnight before being inoculated with field isolates of ASFV (L'60, Tengani, and DR-II). After 18 hours incubation at 37 C, cytospin smears from the mixture of the floating cells and plastic adhering cells (scraped off from the plastic

98

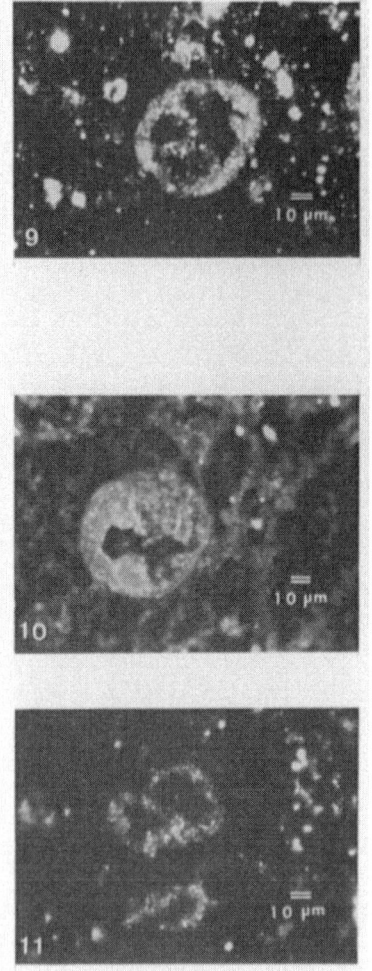

Fig. 9. Vascular pattern of fluorescence for ASFV antigens, presumably in degenerated vascular wall and in mural thrombus, seen in tonsil. Notice numerous infected macrophages in the surrounding tissue. Tonsil, frozen section, anti-ASFV conjugate stain. Case #8085. Killed 8 DPI, L'60B6-BC10 virus (10 generations of segregation in BC culture at limiting dilution).

Fig. 10. Deposition of immunoglobulins in the vascular wall and in mural thrombus. Many infected cell debris in the connective tissue were fluoresced also. Tonsil, frozen section, unfixed, washed, OAIg conjugate stain. Case #8083. Killed 8 DPI, L'60B6-BC10 virus.

Fig. 11. Deposition of complement in degenerated vascular walls and in mural thrombi in a pattern mimic to that with ASFV antigen deposition, as seen in Fig. 9. Tonsil, frozen section, unfixed, RAC conjugate stain. Case #8085.

wall) were stained with ASF antibody conjugate; results of careful observations indicated that the only infected cells were monocytes. In no case could we find infected lymphocytes or neutrophils.

Leukocytes from infected blood. The heparinized blood obtained from pigs during the acute phase of infection with ASFV (L'60, Tengani, and DR-II) were centrifuged in F-D gradients, washed with Medium D, and cytospin preparations were made and stained with ASF antibody conjugate. In each case, infected cells were only monocytes. In several occasions, we observed large, infected monocytes with fragmented nuclei (or phagocytized fragmented nuclei), appearing similar to neutrophilic nuclei, in the bottom fraction; such cells in the bottom fraction were presumably of dead or injured monocytes. In no case could we observe the presence of infected lymphocytes or neutrophils.

Eosinophils. The eosinophilic granules from normal pigs generally autofluoresced with faintly pink to silver white color. However, on other occasions, closely packed granules of uniform size fluoresced with an intense green in frozen sections of lymphoid organs of normal and infected pigs that required some experience to distinguish from specific fluorescence. Conjugation of F(ab)$_2$ of immunoglobulins with fluorescein did not remove this non-specific fluorescence.

Hepatocytes. Hepatocytes were reported to be susceptible to infection (16); however, we could not substantiate this finding in our studies. In only few cases did we observe aggregates of infected large cells in the hepatic acini (presumably foci of extramedullary hematopoiesis in young pigs). The presence of microthrombi with ASFV antigens was not infrequent in blood sinusoids.

Thymus atrophy

It was noted at necropsy that a marked atrophy of the thymus appeared to accompany those cases which died from chronic ASF. Because the thymus is known to be the central organ for proliferation, maturation, and the source of T lymphocytes for the peripheral lymphoid organs, the infection and destruction of REC which provide microenvironment for maturation of T lymphocytes might interfere with this maturation process. Observations were few and comparison with proper control groups were not always available; however, seventy seven of age-matched pigs exposed to a vaccine virus (L'60BM89) alone or vaccinated and challenged with homologous virulent virus (L'60) were investigated, and the results are summarized as follows (Fig 12):

Thymus cell numbers after vaccination only. In pigs killed 18 to 21 days after receiving vaccine virus (L'60BM89), the number of thymus cells were within normal limits as determined by comparison with that of normal pigs sacrificed on the same day. At 28 DPI the cell numbers were slightly less than normal, but thereafter the cell number in the vaccinated pigs increased, and by 42 DPI they averaged about 40% more than normal.

Thymus cell numbers following vaccination and challenge with virulent virus. In vaccinated pigs challenged at 18 DPI with homologous

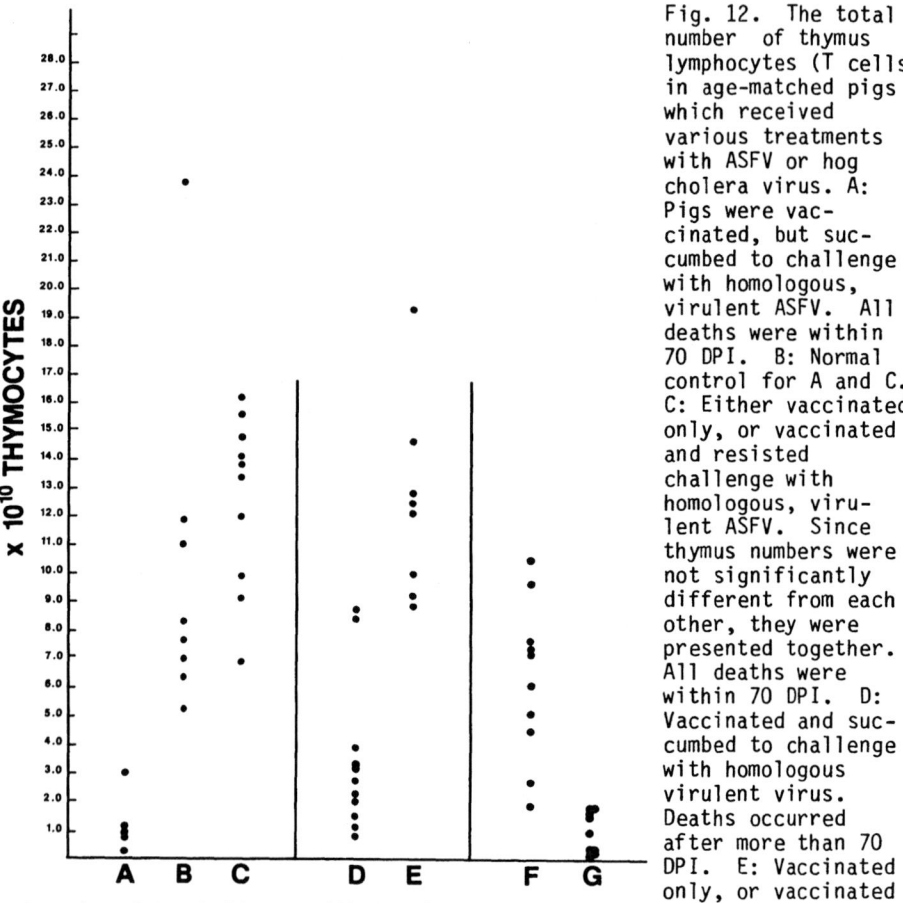

Fig. 12. The total number of thymus lymphocytes (T cells) in age-matched pigs which received various treatments with ASFV or hog cholera virus. A: Pigs were vaccinated, but succumbed to challenge with homologous, virulent ASFV. All deaths were within 70 DPI. B: Normal control for A and C. C: Either vaccinated only, or vaccinated and resisted challenge with homologous, virulent ASFV. Since thymus numbers were not significantly different from each other, they were presented together. All deaths were within 70 DPI. D: Vaccinated and succumbed to challenge with homologous virulent virus. Deaths occurred after more than 70 DPI. E: Vaccinated only, or vaccinated and resisted to challenge with homologous, virulent ASFV. F: Acute ASF. All died within 10 DPI. G: Acute hog cholera. All died within 13 DPI.

virulent virus, there was no significant change in cell numbers within the first 5 days after challenge (DPC). However, during 7 to 21 DPC significant cell losses occurred in those pigs which had clinical signs of ASF. The average was about 50% of normal, and in some cases was reduced to 10% of normal. At 70 DPI the differences in cell numbers between the vaccinated and vaccinated and challenged groups were quite obvious. The challenged group had about 20% of the number of cells present in the group that was vaccinated only. Thereafter, cell numbers in the challenged group increased sharply. At 115 DPI, they were double the number recorded at 100 DPI (Fig. 12).

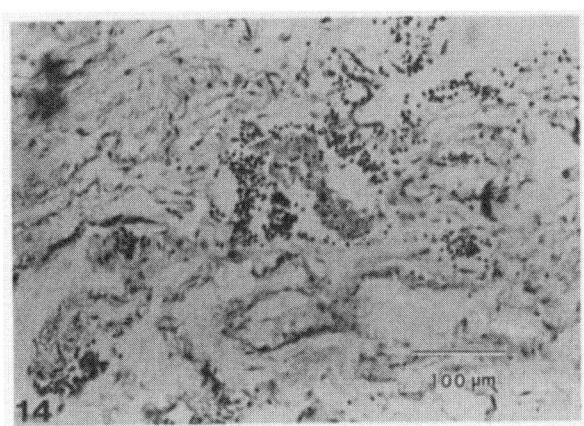

Fig. 13. Marked atrophy of cortex and medulla with condensed appearence of Hassal's corpuscles and reticular epithelial cells. Thymus, paraffin section, 6 μm thick, H. & E. stain.

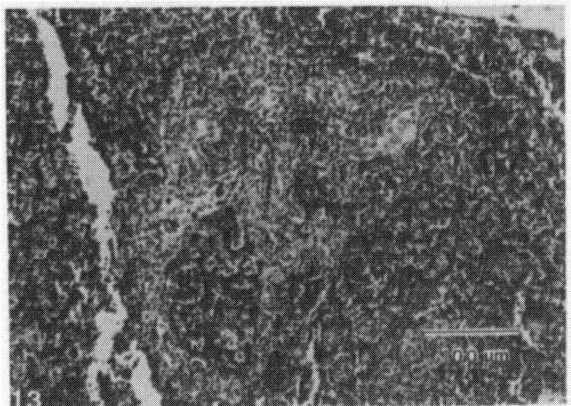

Fig. 14. Extreme atrophy of thymus. Fibrous connective tissue with sparse lymphoid cells are the remnant of thymus tissue. Thymus, paraffin section, 6 μm thick, H. & E. stain.

In intermediate stages of atrophy, the thymus cortex was generally thin and deprived of thymocytes (Fig. 13). In IF studies, infected REC was observed only in atrophic thymus of vaccinated pigs which had succumbed to challenge with the virulent virus. In extreme cases, only fibrous stroma was left in the atrophied thymus (Fig. 14).

Rapid Turnover of Thymus Cells

Three groups of 5 pigs each were housed separately and were treated as follows: (a) vaccinated with L'60BM89BC1 virus $10^{3.0}$ HAD_{50}; (b) vaccinated and challenged at 18DPI with L'60B5 virus ($10^{5.0}$ LD_{50}); (c) normal controls.

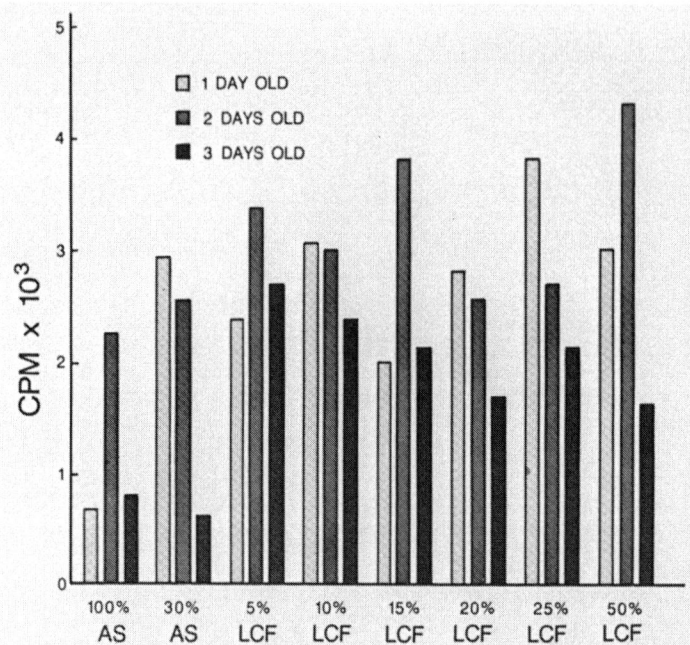

Fig. 15. Determination of optimal concentration of LCF for stimulation of MNC. Plastic-adhering MNC were cultured in Media A containing from 0 to 50% LCF, and in 100% autologous serum as a control. [³H]thymidine (7.5 μCi/ml, 20 Ci/mmole) was incorporated to media. Radio-labeled-adherent MNC were scraped-off at specified time interval for counting radio activity. The maximum DNA synthesis or cell division occurred at 1 to 3 days of culturing. Some cell division also occurred with cells cultured in autologous serum, and considerable cell division was observed when cells were cultured in F15 containing 30% of autologous serum (AS). Although all concentrations of LCF stimulated cell division, 15% LCF was chosen for other experiments mainly for economic reason.

One pig from each group was killed each day, from 3 through 7 DPC. Thymuses were removed and the total weight of each was determined. Counts were made of the cells teased from two weighed tissue samples from each thymus, and the total number of cells per thymus was calculated.

The average number of cells per normal thymus was 124.1 x 10⁹. Using this figure as 100%, the thymus cell counts from the vaccinated, and the vaccinated and challenged animals were expressed as percentages of the normal. The following findings were obtained:

a. Vaccinated only: Starting at 126% of normal at 21 DPI, the counts returned linearly to normal (about 98%) by 25 DPI.

b. Vaccinated and challenged: The thymus cell count was 64% of normal at 3 DPC (21 DPI) and decreased to 58% of normal at 4 DPC, but returned to approximately normal at 5 DPC. At 7 DPC (25 DPI) the count was about 30% above normal.

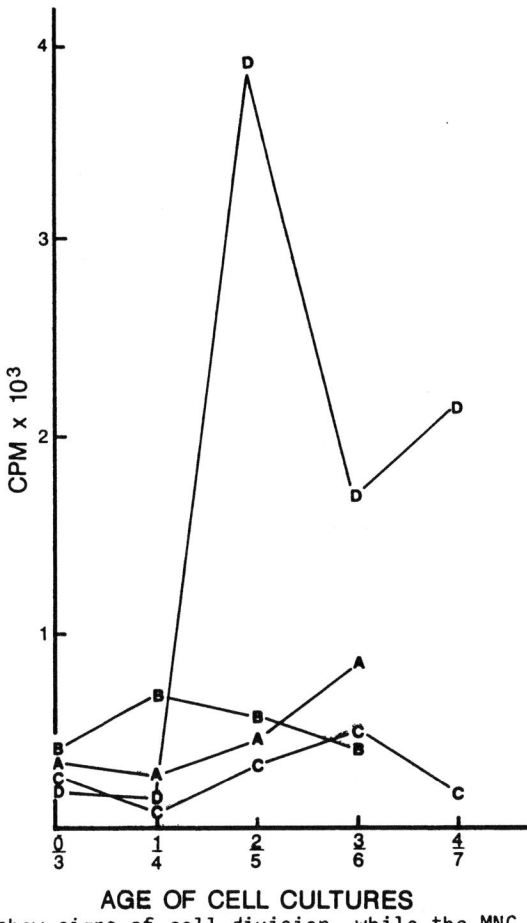

AGE OF CELL CULTURES

Fig. 16. Response of 3-day old MNC to MSF. MNC had been cultured for 3 days in the respective medium (Group A: Medium A; Group B: Medium B; Group C: 30% heat inactivated FBS in F15; Group D: 15% LCF in F15 containing 30% heat inactivated FBS) before replaced with new media of following composition (Groups A: 30% heterologous swine serum in F15; Group B: 15% LCF in F15 containing 30% heterologous swine serum; Group C: 30% heat inactivated FBS in F15; Group D: 15% LCF in F15 containing 30% heat inactivated FBS). Microscopically, MNC in Groups A and B grew very well, and less cell numbers were observed in Group C and D, at 3rd day of culturing in the initial media. After changed to new media, very little DNA synthesis occurred in Groups A and B in 3 days of culturing; however, microscopically, MNC cultures in both media were almost confluent. After changed to new media, Group C did not show signs of cell division, while the MNC in Group D responded to MSF well, and peaked at 2nd day after medium change.

Plastic-Adhering MNC

Since Malmquist and Hay (1) reported that the porcine BC cell culture supported the replication of ASFV in 1960, subsequent reports by others all agreed that blood monocytes were the cells susceptible to ASFV infection. With the exception of monoblasts or promonocytes, the monocytes in the peripheral blood have been considered as terminal cells which are incapable of cell division. Therefore, we have investigated macrophage colony—stimulating factor (MSF) (24) to determine: (a) if it is capable of stimulating cell division of porcine blood monocytes or macrophages, and (b) if stimulated monocytes or macrophages are susceptible to ASFV infection. The results are as follows:

104

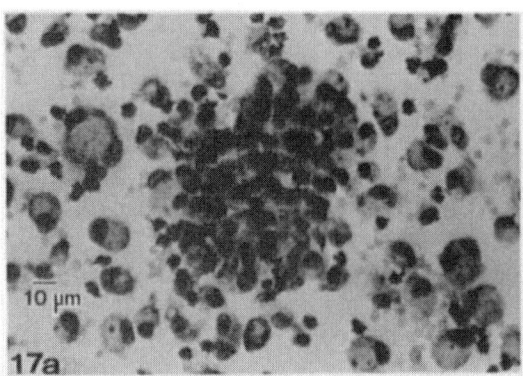

Fig. 17a.
Macrophage
colony formed
by MSF stimu-
lation.
Clusters of
macrophage
colonies of
various sizes
were formed
at about 5
days of
culturing.
The colonies
were very
easy to
detach from
plastic wall
or from giant
cells at the
time of media
change.
Seven-day old
culture.
Giemsa stain.

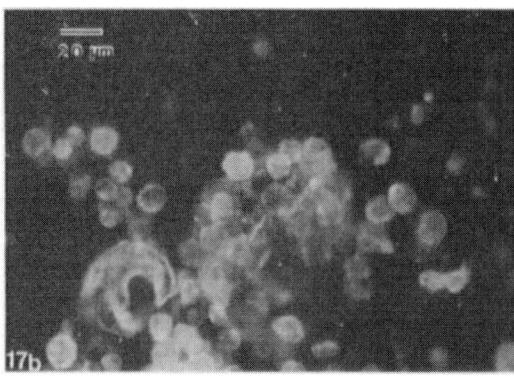

Fig. 17b.
Macrophages
in a colony
stimulated
with MSF were
fully suscep-
tible to the
virus infec-
tion. The BC
cell culture
was cultured
on a slide
with Medium
B, and the
field isolate of Tengani virus was used to infect cells. Harvested 18
hours after infection. Fixed with methanol and stained with anti-ASFV
conjugate.

Stimulation of mononuclear cells. The young, plastic-adhering,
mononuclear cells were stimulated by MSF when 5 to 50% LCF had been
incorporated in culture medium (Fig 15).

Cell division occurred but to a lesser extent by simply culturing
cells in Medium A, or C without MSF (Fig. 15).

The maximum cell division in media containing various concentrations
of MSF occurred in 1 or 2 days of culturing at 37 C (Fig. 15).

The older cultures, which had been incubated in 30% FBS in F15 for 3 to 4 days, responded well to MSF. The MNC cells which became an almost confluent monolayer culture in response to the previous stimulation by MSF, did not respond to the second stimultion by MSF (Fig. 16).

The autologous serum can be substituted by heat inactivated FBS to supplement Media B; however, FBS alone, without MSF, did not have a stimulatory effect (Fig. 16). Detachment of adhering cells occurred between 3 and 4 days of culturing.

Near confluent monolayer cell cultures can be obtained in 3 to 4 days of culturing in Medium B, depending on the initial numbers of MNC present; however, cells never fused (except multinucleated giant cells) to form a true confluent cell monolayer, as seen in kidney cell lines.

Morphologic heterogeneity of the stimulated cells. The plastic-adhering populations of MNC, obtained either from BC or from the interface of blood plasma and a F-D gradient, after centrifugation of heparinized blood, contained not only typical blood monocytes but also spindle cells, flat polyhedral cells, and small spherical cells with round nuclei with scanty cytoplasm (mimic to small lymphocytes) often possessed one or two hairy projections in one pole. The latter population has possibly been mistaken for lymphocytes in the past; the same cell type grew into flat, elongated cells later when stimulated by MSF. All cell populations which were stimulated by MSF possessed endogenous peroxidase and latex phagocytising activities. Upon stimulation by MSF, the large, spherical, highly refractile cell population (presumably monocytes) was the only population forming colonies within 5 days of culturing with MSF (Fig. 17a) and pre-ferentially grew on top of multinucleated giant cells; frequently, the latter became the prominent population at this stage of culturing. The macrophages in the colony are fully susceptible to ASFV infection (Fig. 17b), and the structural viral antigens (9) were demonstrated by the indirect IF when the swine antiserum raised against inactivated, purified virus was used as the first serum.

Susceptibility of the stimulated MNC to ASFV infection

The stimulated cells (cultured in Media A or B) and unstimulated cells (cultured in Medium C) were grown in T_{25} culture flasks. They were subsequently inoculated with a field isolate (Tengani) of ASFV. The MNC cultures at 1 to 9 days of age were washed three times with

Medium D, and one ml each of ASFV containing approximately $10^{5.0}$ HAD_{50}/ml was inoculated. After 1 hour adsorption at 33 C with redistributions of the inoculum at 10-minute intervals, the inoculum was removed by aspiration, and the cell cultures were washed 3 times with Medium D and were either overlayed with 7 ml of fresh, respective culture media (containing 0.2% of autologous erythrocytes), or replaced entirely with Medium C. The infected cultures were incubated at 33 C and observed for hemadsorption and CPE. All culture fluids were harvested 8 DPI and were stored at -70 C until assayed by HAd test.

The results indicated that: (a) MNC cultured for up to 8 days of age in Medium C were susceptible to ASFV infection; (b) MNC cultured in Medium A or B were susceptible to virus infection up to 3 days of age, but a progressive decline in virus yield was observed when the

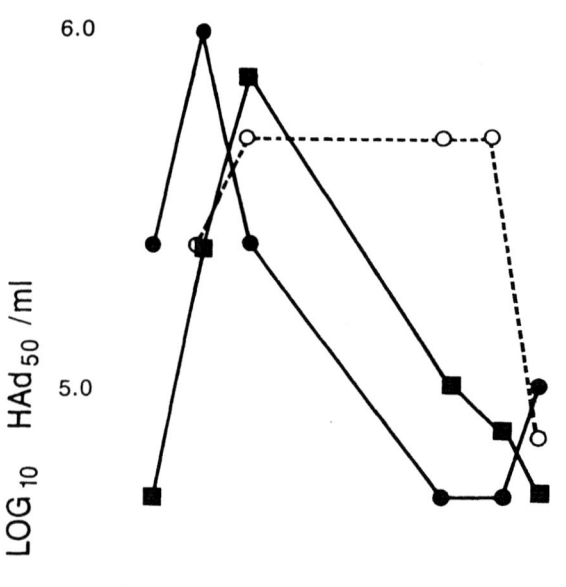

Fig. 18. Susceptibility of cells stimulated by MSF to ASFV infection. The MNC cultures inoculated were cultured 1 to 9 days in the respective medium. All cultures were inoculated with Tengani virus, and virus produced was harvested 6 DPI. The control MNC cultured in Medium C (○--○) were equally susceptible to ASFV infection up to 8 days of age. The MNC cultured in Medium A (●—●) and Medium B (■—■) were susceptible up to 3 days of age; however, both cell groups progressively lost susceptibility, thereafter. Morphologically, both cell groups appeared healthy, and HAd cells were sparsely present up to 8 DPI (end of observation); very little evidence of infection was present.

corresponding fresh culture medium was restored after inoculation (Fig 18); (c) the virus yield by both stimulated and control cultures were comparable up to 8 days when Medium C was used for overlaying the cultures after virus inoculation (Table 2).

Table 2. Yield of ASFV by Mononuclear Cells Stimulated with MSF.

	Age of cells (days)				
	4	5	6	7	8
Test Group	5.42*± 0.14	5.21 ± 0.16	5.09 ± 0.16	5.08 ± 0.15	4.96 ± 0.17
Control Group	5.50 ± 0.13	5.25 ± 0.16	5.00 ± 0.16	5.04 ± 0.16	5.30 ± 0.19

Cultured cells were inoculated with 1 ml of $10^{6.0}$ HAD_{50}/ml of Tengani isolate at specified age of cells. After 1 hr adsorption at 33 C, the cultures were washed with Medium D, and all cultures were overlayed with Medium C.
*log_{10} HAD_{50}/ml at 95% confidence interval.

The positive HAd reaction

In an ASF virus-infected culture prepared for examination, varying numbers of leukocytes will be seen with erythrocytes attached to their surfaces, forming rosette-like patterns. The entire surface of the leukocyte may be covered, presenting a rasberry-like appearance. Large numbers of erythrocytes may be absorbed several layers deep around a single infected leukocyte. The red cells are quite firmly attached and can stand rather vigorous agitation without dislodging. In fact, the infected leukocytes may become detached from the glass or plastic wall and may be seen floating across the field with erythrocytes still attached to its surface. The infected cells with adsorbed erythrocytes detach from glass or plastic walls within 24 hours after infection. Therefore, a more strongly positive hemadsorption reaction is seen with BC culture in plastic plates than with Leighton tubes, because detached and undetached HAd positive cells are both present together at the bottom of the plastic plates, where they may be detected with an inverted microscope, while detached HAd positive cells are generally not present to be detected in inverted Leighton tubes. Although the number of leukocytes displaying HAd is indicative of the amount of virus present (one MNC produce approximately 1,000 HAd_{50} units in BC culture), a single cell displaying typical HAd is sufficient to declare the culture

positive. Usually, if such a culture is held another day, the infection
will spread, and more cells will be involved in the HAd reaction. On
certain occasions, however, for reasons not fully understood, the number
of hemadsorbing cells did not increase after an extended period of
observation. A period of 8 to 10 days of observation is required when
no HAd nor detectable CPE are observed. After cell lysis, all infected
cells become amorphous cell debris without any evidence of HAd.

The negative HAd reaction with or without CPE

The inoculum prepared from the chronically infected animal
generally contains antibody titers high enough to inhibit the HAd
reaction, even though the leukocytes are infected with ASFV.
Furthermore, the non-hemadsorbing virus (NHV) does not induce HAd, but
only CPE. Under these circumstances, 0.1 ml of the culture content is
removed, and a cytospin preparation is made (1,000 rpm, 3 minutes). The
specimens are air-dried, fixed with methanol (2 minutes) or acetone (5
minutes) at room temperature, stained with the anti-ASF conjugate and
examined with a fluorescence microscope. In either case, positive IF
indicates the presence of virus. When the result of IF test is negative,
one or more subpassages to new BC may be required before the presence of
ASFV becomes apparent.

Non-specific observations

In cultures more than 48 hours old, macrophages are seen that have
phagocytized large numbers of erythrocytes. Such cells may super-
ficially resemble a positive HAd reaction. On close examination, the
erythrocytes are clearly seen to be lying within the cytoplasmic
membrane of the macrophage.

At other times, for reasons not fully understood, red cells may
form clusters or crenate. Some of these clusters may become trapped
among the leukocytes and somewhat resemble HAd. However, the clusters
are usually more irregular in form, are more easily dispersed, and, with
a little more agitation, will become dislodged from the leukocytes.

Detection of hemadsorbing cells in the blood drawn from pigs acutely
infected with ASF virus

From limited experience, the hemadsorbing cells in BC culture, made
from acutely infected pigs, are detected within few hours after preparation.
The frequency of observing HAd-positive cells seems to depend upon the
stage of infection that the pigs are in. Samples taken from the height

of their fever response to the moribund stage were all positive for the
HAd reaction within 3 hours of incubation at 37 C. The HAd positive cells
are either attached to the glass wall or floating in the culture medium.
In parallel studies with IF test where cytospin specimens, either prepared directly from buffy coat or the MNC prepared from F-D centrifugation
of heparinized blood of acutely infected pigs, were stained with
anti-ASFV conjugate and examined for IF, both methods detected the same
numbers of positives. In more detailed studies with the latter method
(F-D centrifugation), many IF-positive cells (dead monocytes) were found
in the bottom fraction of F-D centrifugation preparation.

HAD_{50} or $TCID_{50}$ titration in BC cultures, and virulence and infectivity
titration in pigs

The HAD_{50} or $TCID_{50}$ units were used to express the virus titer
obtained in titrations in BC cultures; the former was used for HAd virus
and the latter for NHV. We adopted the inoculum size of 1.0 ml /
Leighton tube in which results were expressed HAD_{50}/ml without correction of titer. This method is recommended, because: (a) There is no
valid conversion factors for expressing the results of titrations in
HAd_{50}/ml when various inoculum sizes are being used in the experiments;
(b) The inoculated BC cultures without infection generally retain their
morphology better than the uninoculated controls at or beyond the 7th
day when the test is usually not yet completed; (c) The BC culture is the
only method of titrating field isolates which have not been adapted to
cell lines. In this regard, we found that the virus titer expressed by
HAD_{50} or $TCID_{50}$ units is extremely useful in studies of virulence and
diversity of virus population of an isolate in pigs (10,20). The
results of experiments led to a characterization of virulence of ASFV as
follows: (a) Highly virulent virus can induce 1 PID_{50} and 1 LD_{50} in
pigs with $\leq$10 virus units, and the pigs which had clinical signs of ASF
always died with acute ASF. (b) Moderately virulent virus can induce
clinical signs of ASF in pigs with $\leq$10 virus units, but it requires 20
to 562 virus units to induce 1 LD_{50} in pigs. Immunological deaths and
recovery from the clinical disease were not uncommon. (c) Virus of low
virulence requires 56 to 10,000 virus units to induce clinical signs (1
PID_{50}), and it requires 56,200 to 3,120,000 virus units to induce an
acute death (1 LD_{50}) in a pig. Titrations of virus in BC cultures and
pigs were highly reproducible (10).

Determination of inoculum size for different culture vessels

In titrating a sample with BC cultures in 24-well plates, the inoculum size was determined by the assumption that the efficiency of virus "take" is directly proportional to the depth of the inoculum. Therefore, the proportionate inoculum size for a well (2 cm^2) in 24-well plate is 2/5 of that of Leighton tube (5 cm^2) which is 1.0 ml/tube; hence, 0.4 ml inoculum size for a well of 24-well plate was determined. The results of simultaneous titrations of a sample in the two different culture vessels were almost identical (Table 3).

Table 3. Comparison of virus titer (HAD_{50}/ml) obtained with different culture vessels.

Culture vessels	Inoculum size	Virus dilutions				Virus titer (HAD_{50}/ml)
		Virus = L'60BC6-1				
		10^{-6}	10^{-7}	10^{-8}	10^{-9}	
24-well plate	0.4 ml/well	0/8	0/8	5/8*	8/8	$10^{7.88} \pm 0.18$
Leighton tube	1.0 ml/tube	0/8	0/8	5/8*	7/8	$10^{7.88} \pm 0.18$
		Virus = L'60B6				
		10^{-6}	10^{-7}	10^{-8}	10^{-9}	
24-well plate	0.4 ml/well	0/16	0/16	8/16	15/16	$10^{8.06} \pm 0.14$
48-well plate	0.2 ml/well	0/16	0/16	10/16	14/16	$10^{8.00} \pm 0.15$

5/8* The numerator is numbers of wells or tubes devoid of infection. The denominator is numbers of wells or tubes used for each dilution.

Likewise, 0.2 ml/well of inoculum size was determined for 48-well plates where 24-well plates were used for comparison. The titration results using spleen suspension of L'60 isolate were almost identical (Table 3). Therefore, all results were recorded without correction and expressed in HAD_{50}/ml. For practical reasons, the following numbers of tubes or wells are used for each dilution of the test samples: 8 tubes for Leighton tubes, 8 wells for 24-well plate, and 16 wells for 48-well plate.

Detection and isolation of NHV from viremic blood or organs by BC culture

A 10 or 20% suspension of sonicated, defibrinated blood or organs has been used as inoculum in BC cultures for detection of viremia or for isolation of virus. The positive HAd indicates the presence of ASFV in the inoculum, and the isolated virus will likely be called a HAV even though the HAV population might constitute a minor population in the mixture with a NHV population. Thus, the presence of NHV will escape detection. Therefore, we routinely dilute the inoculum from 10^{-1} up to 10^{-9} by 10-fold dilutions and inoculated 4 wells/dilution in BC culture in 24-well plate. The 24-well plate is recommended because the NHA—positive well, which has been identified by IF test with cytospin preparation (0.1 ml), still contains enough virus (0.7 ml) for propagation or storage of the virus in a freezer at -70 C (the virus cultured in BC culture can not be kept at 5 C, because it usually loses 90% of its titer after a week storage). In this way, we have detected the presence of NHA in the test samples where it previously remained undetected (20).

In this regard, we have attempted cloning virus through 20 generations of passages in BC culture at limiting dilution. In one instance, L'60BM89BC20-2 virus showed a typical hemadsorption reaction in BC culture; however, it was determined later that the virus isolated contained not only HAV but also NHV and atypical HAV populations (20).

Stability of the non-hemadsorbing nature and virulence of cloned virus by plaque purification

The two NHV cloned by plaque purification never regained hemadsorbing ability and did not change in virulence in pigs under various test conditions (10,20). An additional test for stability of the non-hemadsorbing characteristic was made as follows: A high titer inoculum (approximately $10^{7.5}$ PFU/ml) prepared from a NHV clone (HT-NHV-8111), plaque purified 4 times from the parent virus (HT-NHV-P6), was inoculated to 19 roller bottles (containing approximately $2 \times 10^{8.0}$ Vero cells/bottle). After 6 days of culture at 33 C, CPE was complete, and cell debris was collected by centrifugation at 1200 g for 30 minutes at 5 C. The pellet was suspended in 200 ml of the culture fluid; it was sonicated, and the supernatant after centrifugation was used as inoculum for the next inoculation of 19 bottles of new Vero cell cultures. Each harvest was checked for the presence of HAV by inoculating BC cultures with serial 10-fold dilutions of the inoculum. All cultures tested up to 18 generations of serial passages were negative for HAd indicating that the nonhemadsorbing characteristic was stable.

Susceptibility of cell lines to field isolates of ASFV

Invariably, all cell lines tested were found infected when tested by direct IF test at 7 hours postinoculation (HPI). The virus antigens fluoresced as fine stipplings of a uniform size, scattered in the cytoplasm; however, there was no intranuclear fluorescence nor fluorescence for inclusion bodies (site for virus DNA synthesis and virus assembly) discernible. At 24 HPI, the majority of infected cells did not change from the degree of cytoplasmic fluorescence seen at 7 HPI. Only rarely seen were infected cells with viral inclusion bodies together with strongly fluorescing stippling in the cytoplasm. Attempts to demonstrate an increase in virus titer at 24 HPI compared to 1 HPI failed.

After adsorption of the field isolate of ASFV for one hour at 37 C, a Vero cell culture was overlayed with agarose (0.7% agarose in Medium E). When the agarose solidified, the culture was incubated at 33 C for 7 days. No plaques or CPE were visible under microscope by this time. However, extremely small microscopic plaques (consisting of 3 to 7 infected cells) were visible after staining by the IIPS method. Previously, the PFU titer was found to be 0.94 $\log_{10}$ lower than that by HAD_{50} titration in BC (9). When using this relationship for comparing the results obtained by a parallel titrations in BC and Vero cell cultures, it was found that approximately 60% of virus population had ability to infect and produce virus, and to spread to neighboring cells forming plaques (data not shown).

Presence of defective interfering (DI) particles in spleen of pigs infected with field isolates of ASFV

The CPE of the infected Vero cell sheet at 3rd or 4th passage of field isolates is generally completed by 6 DPI. However, in two virus isolates (Tengani and Uganda), the cell sheet destruction did not progress beyond 50% of total cells; frequent changes of fresh culture medium or repeated virus passages in cell cultures did not enhance destruction of the cell sheet. Nevertheless, these cells were infected with the virus, as evidenced in direct IF test; therefore, it is assumed that DI particles (25,26,27) may have played a significant role in inducing this phenomenon.

Production of high titered virus preparation for virus purification and for diagnostic antigen

A virus titer of approximately $10^{5.0-6.0}$ PFU/ml is generally obtained with the culture fluid when a relatively low titered virus preparation is used as an inoculum. Nontheless, the total virus titer

in the culture fluid is almost identical to that in the cell debris.
Therefore, it is our usual practice to infect a T_{75} flask Vero cell
culture with 7 ml of culture fluid from the previous passage and
incubate for 6 hours at 37 C with occasional agitation (a higher "virus
take" occurs at 37 C than at lower temperatures). The inoculum is
removed, and the culture washed three times with 10 ml each of Medium D
before 15 ml of Medium D are overlayed. The CPE usually was apparent
within 24 hours. After 4 to 5 days of incubation, when CPE was
complete, the cell debris suspended in the culture fluid was chilled to
5 C and kept in an ice bath during sonication (six 10 sec cycles of
sonication at the maximum output and 10 sec of resting). The sonicate
was centrifuged 800 xg for 30 minutes at 5 C, and 7 ml each of the
supernatant were used to infect fresh Vero cell cultures grown in two
T_{75} culture bottles. The inoculated bottles were treated as before. In
general, CPE is complete in 3 days.

After decanting the culture medium, a roller bottle (approximately
2×10^8 Vero cells) was inoculated with 20 ml of the inoculum prepared
from the last virus passage. The inoculated bottle was rolled for 4
hours at 33 C for virus adsorption. The inoculum was removed, and the
infected cell sheet was washed once with 50 ml of Medium D. The culture
was then overlayed with 50 ml of Medium D and incubation continued.
After 3 days of incubation at 33 C, a 50-ml inoculum was prepared as
before. Four fresh Vero cell cultures in roller bottles were inoculated
with 10 ml each of inoculum, prepared as before. The process was
repeated expanding the scale of production until a large stock of high
titered ($10^{7.5-8.0}$ PFU/ml) virus was obtained. The virus preparation
for virus purification and antigen production was prepared from this
virus stock to ensure a consistently high titered virus yield.

<u>Production of field virus in alveolar macrophage cultures</u>

When pigs were obtained from a clean herd, devoid of pneumonia, the
alveolar macrophage cultures were extremely useful for obtaining large
numbers of spontaneously susceptible cell cultures (28). Admittedly,
our experience was limited and results varied from time to time, but the
following findings were typical for the alveolar macrophage cultures
prepared here. When one liter of Medium D containing 10 x antibiotics
was used in alveolar lavage, about 400 ml could be recovered from the
lung. Approximately, $2 \times 10^{9.0}$ cells of plastic adhering alveolar cells/

lungs were obtained from a young pig of approximately 50 Kg body weight. The numbers of plastic adhering MNC from defibrinated blood, obtained at the time of sacrifice (1.2 l), were approximately 1/5 of that by alveolar lavage. After overnight culturing (with 10 x antibiotics present) at 37 C, unattached cells were removed by washing with Medium D, and the culture was ready for inoculation. The alveolar macrophages attached to plastic walls and grew very well, forming an almost confluent monolayer culture overnight when the seed cell numbers were adjusted to 7×10^5 cells/cm^2. The virus yield was approximately the same as that obtained by BC cultures, e.g. approximately 1,000 HAD_{50}/cell.

In instances where the culture was contaminated by mycoplasma, the yield of virus was extremely low even though abundant numbers of macrophages were available for infection.

Microplaque assay (MPA) by immunoperoxidase (IPRO) technique and conventional plaque assay (CPA)

After 5 to 7 DPI, plaques were visible in monolayer cultures of Vero cells that had been inoculated with the uncloned, Vero cell adapted L'60 virus. About 70% of the plaques were microscopic to pinpoint in size and the rest were large plaques (approximately 1.2 mm in diameter). For practical purposes, the inoculum size of 0.1 ml and adsorption time of 60 minutes at 37 C was found adequate for 6-well plate (10 cm^2/well). In cultures stained by the IPO technique (9), tiny brown plaques were visible under 10 times magnification at 2 DPI. At 3 DPI, sharply defined, dark-brown plaques, 0.1 to 0.3 mm in diameter, were seen in MPA. At this stage, only a fraction of plaque counts (approximately 20%) were visible (pin-point size) with CPA; therefore, the MPA is superior for detecting early plaques. The results of plaque counting statistics indicated that the virus titer obtained at 3 DPI through 7 DPI by MPA were comparable (9). A linear relationship between viral concentration in the inoculum and plaque numbers was observed. The maximum limit in plaque count was found to be 500 by MPA at 3 DPI and 200 by CPA at 7 DPI. Viral titers obtained by both MPA and CPA were comparable, and both methods were reproducible and reliable. The titer obtained by HAd test (HAD_{50}/ml) was 0.94 log higher than the titer with MPA (PFU/ml).

IIPS Test for Antibody Detection. We have developed the IIPS test as a fourth generation test for detecting antibody against ASFV; it can

be used as a screening test as well as the final test (6). It has all
the desirable features of IIF but is more efficient and convenient (6).
In addition, the results can be read with the unaided eye under ordinary
light, making it possible for a single technician to conduct 400 test in
a day. Thus far, we have evaluated the method for the ability to detect
antibody to ASFV by comparing it with the IIF, ELISA, and IEOP tests
(6). The results clearly indicated that IIPS and IIF tests are far
superior to ELISA and IEOP tests in sensitivity and specificity. Subsequently, we have attempted to substitute OAIg antibody with Protein A
in IIP test without success. The protein A-HRPO conjugate only detected
13% of cases identified as positive by either IIPS or IIF tests on sera
from blood taken within 10 days after exposure to ASFV. Likewise, the
detection rate for cases beyond 10 days of exposure was about 56% of
that by other tests (Table 4).

Table 4. Low sensitivity of antibody detection by Protein A-HRPO
conjugate.

DPI	Tests			Positive* cases	Test cases
	IIPS	IIF	Protein A		
0	0	0	0	0	38
≤ 10	14 (93.3%)	15 (100%)	2 (13.3%)	15	95
$\geq 10 \leq 14$	48 (93.3%)	47 (97.%)	24 (50.0%)	48	65
> 14	52 (100%)	52 (100%)	32 (61.5%)	52	56

*Numbers of cases positive by IIPS or IFI test or both.

Hypergammaglobulinemia. The occurrence of hypergammaglobulinemia
is a reflection of increased rather than impaired B lymphocyte function
and is due to a continuous stimulation of antibody production by viral
antigens (13). It was observed in all pigs acquiring ASF pneumonia,
regardless of the virulence of the virus used (Pan, I.C., unpublished
data). The increased gammaglobulin in the early phase of the disease
(14 to 21 DPI) were electrophoretically oligoclonal (IgG2 and IgM) in
nature, and became polyclonal later (Fig. 19). On certain occasions,
when an infected pig had a recurrence of fever after a period of clinical
recovery, a sharp, almost myeloma-like band, appeared in IgG_2 fraction
(Fig. 19). In a few cases, the increase in the IgM fraction continued
up to 63 DPI without significant increase in the IgG fraction (data not

shown). The increased Ig contained antibody activities, as shown in a reverse immunoelectrophoresis (Fig. 20).

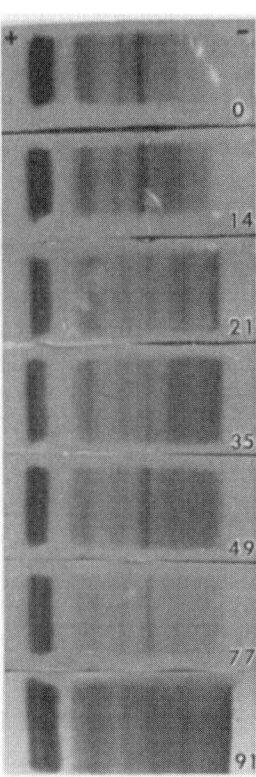

Fig. 19. Electrophoregram of swine sera obtained from sequential bleedings from a swine exposed to L'60-BM89 virus and became chronic ASF. Samples were electrophoresed on cellulose acetate and proteins stained with ponceau S. The number at the edge of each electrophoregram indicated DPI that the blood sample was drawn. The first sign of increase in immunoglobulins was in IgM (beta 2) fraction 14 DPI, and the increase persisted to 63 DPI. A sharp increase in slowest migrating IgG (almost a mycloma protein-like) was observed 21 DPI but soon became more polyclonal in IgG fraction and polyclonal hypergamma-globulinomia 49 DPI. The gamma globulin level returned to normal 77 DPI; however a marked increase in IgG2 was again observed 91 DPI in response to recurring fever. Extensive ASF pneumonia was observed at necropsy 91 DPI.

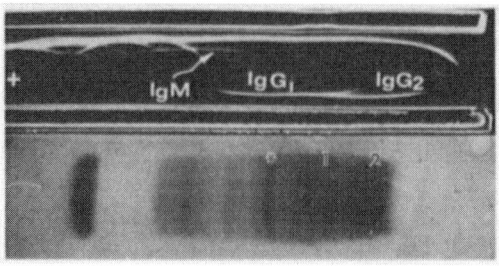

Fig. 20. Reverse immunoelectrophoresis. To identify fractions of Ig containing antibody activities in a hypergammaglobulinemic sreum, a test serum was electrophoresed in agarose (0.7% agarose in a barbital buffer, ionic strength 0.05, pH 8.6; same buffer except ionic strength of 0.1 in buffer reservoirs) for 90 minutes at 250 volts, D.C. Two troughs along the pass of electric currency were cut, and the rabbit serum against swine whole serum was aplied to the top trough and the ASFV soluble antigen (IEOP antigen) applied to the bottom one. The precipitating lines in the lower field indicated that both IgG1 and IgG2 contained major antibody activities. The corresponding electrophoregram at bottom indicated that increased IG were both IgG1 and IgG2 (*transferin; 1, IgG1; 2, IgG2).

DISCUSSION

The results of our studies on susceptible cells by the IF technique on pigs that either died or were killed at the moribund stage of acute ASF generally supported the findings of others (1,5,16-18). Moreover, certain cell populations, which have never been reported previously, were found to be susceptible to ASFV infection. On the other hand claims that certain cell populations were susceptible could not be substantiated in our studies.

The mesangial cells in the glomeruli seem to be infected early in the course of acute ASF, usually 1 to 2 days after the onset of clinical signs of ASF. Subsequently, the presence of cell debris fluorescing for ASFV antigens in the mesangial area, formation of microthrombi containing sandy or granular ASFV antigens which appear obliterate capillary lumens, and ASFV antigens present along capillary walls were found in the glomerular tufts of the kidneys of dying or dead pigs (Figs. 7a,b,c). In addition to the observed changes in other organs and tissues (Figs. 1,8), we feel that dying pigs develop disseminated intravascular coagulation (DIC), perhaps caused by extensive tissue damage and formation of soluble immune complexes. Since it is out of the scope of this chapter to discuss DIC in more detail, readers are referred to a review article by Moore (29).

In only two cases did we find the presence of ASF antigens, without inclusion bodies, in the cytoplasm of renal tubular epithelial cells. It suggests that the infectious viral particles leaked out from glomerular capillaries, or less likely from intertubular capillary beds, and infected tubular epithelial cells. As indicated by the lack of inclusion bodies, virus replication did not occur. This is much like the first passage of a field isolate of ASFV attempted in a kidney cell line.

Lymphocytes were reported to be susceptible to ASFV infection (16,21), and extensive necrosis of lymphoid organs was attributed to lymphocyte infection (30). However, we could not find any evidence indicative of infection of lymphocytes either in T or B dependent areas in lymphoid tissues, or in the blood of infected pigs. Cultured lymphocytes, obtained from normal swine inoculated with ASFV, also failed to show any indication of lymphocyte infection. This negative finding agrees with those reported by others (31,32).

In this regard, Wardley et al (21) reported that lymphocytes were susceptible to ASFV infection, but the same author reported that both T and B lymphocytes were stimulated by ASFV to proliferate in vitro (33). No explanation of these conflicting observations has been offered.

The occurrence of hypergammagloublinemia was included here because it clearly indicates that the humoral immune system is not impaired. The increased Ig level, the shifts from IgM to IgG_2 and from oligoclonal IgG to polyclonal IgG perhaps reflects a widening spectrum of specificities of the B lymphocyte clones (Fig. 19). In this regard, hypergammagloblinemia was found to consistently accompany ASF pneumonia (Pan, I.C., unpublished data) reflecting a constant stimulation by ASFV antigens released from the pneumonic lungs. The unneutralizable nature of specific antibody for ASFV in the absence of C (Pan, I.C., unpublished data) results in the pneumonic lungs becoming a large reservior of ASFV antigens. A shift in virus populations was frequently observed in the pneumonic lungs (20). The sharp increase in IgG_2 in the serum sample of 91 DPI bleeding (Fig. 20) perhaps reflects this shift in virus population.

The lymphocytosis with functions of T and B lymphocytes remaining intact after exposure to a vaccine and a virulent viral challenge has been reported elsewhere (34,35). As cautioned before, one should not literally interprete parameters obtained from in vitro tests as a quantitative measure of the degree of sensitization of the whole body to ASFV antigens where peripheral blood is used as a source of lymphocytes. It merely reflects the level of sensitized lymphocytes in the circulation at the time the blood was drawn; at best, results should be judged on an "all or none" basis, and not on a quantitative basis as reported by others (36).

Neutrophils were reported by others to be susceptible to ASFV infection (16,23). However, their findings, could not be substantiated in this study. The lack of neutrophilic granules in the EM picture of the infected cell (23) excluded the possibility of identifying the cell in question as a neutrophil.

The new finding that REC are susceptible to ASFV infection prompted us to investigate whether thymus atrophy could be attributed to direct destruction by ASFV infection. However, no firm evidence to support this assumption was found in this study. The data suggest that a depletion of thymus cells occurs shortly after exposure to ASFV (either vaccine or virulent); but while surviving the infection, the thymus within 3 weeks becomes hyperplastic, and the normal cell level is restored by 25 DPI. If these animals are challenged at 18 DPI, the thymus cells are again mobilized and the gland reaches its lowest cell count by 4 DPC. But the thymus quickly returns to normal by the following day (5 DPC) and by 7 DPC the thymus is definitely hyperplastic. The quick turnover rate of thymocytes in the group of pigs that were vaccinated and resisted challenge with virulent virus suggests that REC were at least not irreversibly damaged. On the other hand, marked depletion of thymocytes and infected REC in vaccinated pigs, which reacted adversely to challenge with virulent virus, suggests that an enormous mobilization of thymocytes occurred within a very short time and depleted the supply of new thymocytes.

Thymus depletion is also a marked characteristic of hog cholera infection (37), and on the few occasions that observations were made in this laboratory the reaction in acute hog cholera appeared to be even more intense than that occurring with ASF (Fig. 12).

Infection of macrophages in alveolar lumen and alveolar septa was almost a constant finding in edematous and hematomatous areas of the lung of pigs which died from acute ASF infection. In the alveolar septa, the presence of a population of neonatally colonized blood monocytes attached to endothelial cells and functioning like fixed macrophages, was reported for swine (38). The ASFV susceptible nature of alveolar macrophages in pigs with the pneumonic form of ASF has been reported elsewhere (39).

The plastic-adhering MNC were stimulated by coculturing with LCF, and distinct cell proliferation and colony formation were observed by days 2 and 5 of culturing, respectively. In spite of heterogeneous morphology of dividing plastic adhering cells, all dividing cells possessed phagocytic and peroxidase activities. However, the virus yield by stimulated cell cultures was almost constant from 4 to 8 days of age and was similar to that of the control group (Table 2) suggesting a non-virus productive nature for the majority of stimulated cell populations. Despite of this, nearly 100% of the cells were infected, as detected by IF studies. This was rather puzzling because a precaution was taken to avoid interference from biologically active substances, e.g. lymphokines, monokines, and interferon(s), in the culture fluid which might suppress virus infection (Table 2). A tentative conclusion can be drawn from rather limited data that: (a) MSF contained in the LCF stimulates swine MNC to divide and to form colonies; (b) the cell population(s) which are susceptible to virus infection and produce infectious progeny do not respond to MSF stimulation; (c) the majority of those populations which respond to MSF stimulation are susceptible to virus infection, but do not produce infectious progeny; (d) plastic-adhering MNC consist of heterogeneous populations of differing cell lineage.

Since ASFV primarily infects and causes CPE in the macrophage system, the virus may usually be detected in the blood of the acutely sick animal. The simple technique using BC cultures from an acutely infected pig for demonstrating hemadsorbing cells as described here is not intended as a replacement for the regular HAd test. Firstly, it can only be performed with the blood of a live animal. Secondly, the blood sample must be received in the laboratory shortly after it is drawn (less than 4 hours with antibiotics and refrigeration). If these conditions can be met, it can be used to demonstrate HAd within a few hours in acute ASF.

When cell lines were inoculated with field isolates of ASFV and examined with direct IF test, numerous cells were found to be infected, as evidenced by cytoplasmic stipplings without inclusion bodies at 7 HPI, similar to those observed in the earlier stages of infection (3 HPI) in Vero cells with cell-adapted virus as reported previously (40). These same IF findings persisted up to 24 HPI with very few cells possessing the inclusion bodies in which synthesis of viral DNA and virus assembly are known to occur (40,41). Furthermore, there was no convincing evidence from virus titrations that infectious virus was formed at this stage. Nonetheless, the presence of "adapted" virus progeny could be demonstrated after the 3rd blind passages with the method outlined earlier. We feel that the so-called "adaptation" process may well be a virus selection process in which only certain virus populations capable of growing in the particular cells are nurtured in the subsequent blind passages. Therefore, it is possible that populations of "adapted" virus may represent only a fraction of the populations contained in the original inoculum.

This constitutes a serious problem when analysing data obtained from in vitro experiments where viruses of different origins were used;

the spectrum of virus populations contained in the cell-adapted virus may differ greatly from that of the original virus inoculum. Therefore, the data obtained from experiments conducted in vitro can not be directly interpreted as occurring in vivo.

In this regard, any given plaque purified virus clone is invariably virulent to that cell line to which the virus was adapted and to the BC culture as well; the infected cells were readily infected and destroyed. However, the virulence of virus clones in pig is highly variable from clone to clone, ranging from avirulent to highly virulent, and essentially dependent on the virulence factor, which is virus coded (10,20), and the immune reaction of exposed pigs to the virus.

Although hypothetical in nature, extensive destruction of the mononuclear macrophage system may result in the following: (a) massive release of cytoplasmic enzymes in the circulation (47, Pan, I.C., unpublished data) which might initiate the release of toxic biological factors contributing to a worsening of the clinical course; (b) the production of specific antibody, which had been triggered before the extensive destruction of the mononuclear phagocyte system occurred, may form soluble antigen-antibody-C complexes with excessive amount of ASFV antigens (Figs. 9-11); and fragments of complement (C3a, C5a), which have pharmacologic actions similar to those of histamine, are released. The extreme sensitivity of pigs to histamine has been described (43); (c) most of dying or diseased pigs examined so far were bacteriemic indicating a lack of defense against invading bacteria (Pan, I.C., unpublished data); (d) triggering the occurrence of DIC, though a terminal phenomenon in nature, may accelerate the fatal clinical course in acute ASF. Perhaps, synergestic effects of above factors significantly contribute to the acute fatal course that occurs with the virulent virus in pigs; (e) Disrupted chain of events in immune response.

The detection of non-hemadsorbing virus in BC culture with direct IF test is very useful; for, several other viruses of swine, e.g., HC, pseudorabies and swine influenza do infect swine BC cultures with or without CPE. Therefore, differential diagnosis by IF test is extremely useful. The hemadsorption occurs with chicken erythrocytes when BC cultures are infected with swine influenza virus (Pan, I.C. and Buterfield, W.K., unpublished data).

Initially, virus cloning of field isolates was attempted by the "limiting dilution" method in BC cultures. However, we found that this method is not suitable for cloning ASFV, because an ASFV isolate is composed of extremely heterogeneous populations (10,20,44).

The efficiency of virus "take" is directly proportional to the depth of the inoculum in the culture container, as demonstrated in this study. Therefore, adjustment of the virus titer to correct for the volume used is not necessary for 24- or 48-well plates; for, the virus titer obtained in both Leighton tubes and 24- and 48-well plates were comparable, if not identical, with the sizes of inocula used.

For 96-well plates, the inoculum size is 0.05 ml/well. The virus titer is expressed as $HAD_{50}/0.05$ ml or $TCID_{50}/0.05$ ml (Hess, W.R. and Endris, R.G., personal communication). In several reports where the same inoculum size was used, the virus titer was multiplied by a factor of 20, correcting for the inoculum size, to obtain a virus titer per ml (45-47). The virus titers obtained in this way may not be comparable to those obtained with inoculum size of 1 ml/tube for Leighton tube and 0.4 or 0.2 ml/well for 24-or 48- well plate, respectively. For this reason, an appropriate inoculum size for 96-well plates should be sought before the virus titers reported from different laboratories can be meaningfully compared.

Although viral titers obtained by MPA and CPA are comparable, the following advantages favor the use of the MPA: (a) The assay can be completed in 3 DPI; (b) it enables the investigator to detect and quantitate the small plaque-forming viruses that are likely to be missed by the CPA when an uncloned ASF viral population is titrated; (c) plaques are stained specifically for ASF virus, thus the authenticity of a plaque is never in doubt, as it sometimes is when plaques containing undetached cell debris are encountered with a CPA (especially with uncloned virus); and (d) the maximum limit of accurate plaque counting with the MPA far exceeds that of the CPA. The accurate counting of up to 500 plaques in a well of 10 cm^2 is an attractive feature for plaque reduction in neutralization test.

Initially we used a direct staining method for MPA; however, currently we routinely use an indirect method where the staining procedure is almost identical to the IIPS method. An antiserum obtained from swine chronically infected with ASFV is used as the first antibody and HROP conjugated OAIg antibody as the second. With this change in technique, we get a clearer background than by the direct method. Presently, we incubate the infected cell sheet, overlayed with agarose, in an inverted position at 33 C for 4 to 5 days. Stained plaques were large enough to be counted with the unaided eye, and the microplaques, which can not be seen with conventional plaque assay, can be counted under the dissecting microscope.

The advantages of the IIPS test over other tests for antibody detection are numerous: (a) a purified antigen is not required for the test; (b) uninfected cells between plaques serve as a built-in control for detecting antibody against Vero cells that may be present in the test sera; (c) antigen plates preserved at -80 C for 5 years (longest tested) can be used in the test; (d) the results are sharp and clear-cut without so called "gray" area in interpretation. Currently, sera are tested both undiluted and diluted 1:100 with PBS to avoid possible pro-zone effects, thus giving better results than by testing undiluted sera alone. Undoubtedly the IIPS test can be used for any other viruses which can grow in cell lines regardless of whether or not they form visible plaques; applicability for other viral diseases remains to be seen. We tried to substitute OAIg with protein A without success. It appears quite likely that swine IgM and certain subclass(es) of antibody lack affinity for protein A.

More macrophages could be obtained by preinoculation of pigs intra-muscularily with Freund's complete adjuvant (2.5 ml each in two sites) 7 to 10 days prior to sacrifice. However, virus recovery following virus inoculation was very poor. For this reason, pretreatment of pigs with Freund's adjuvant was abandoned. Conversely, the alveolar macrophages from pigs, treated with BCG were reported to support virus production (28). We have no explanation for this discrepancy at present.

If, a pig sacrificed to obtain alveoler macrophage cultures had gross pneumonic lesions, the resulting cultures were not suitable for virus production. In two out of six trials, we have successfully cultured monolayer cultures of alveolar macrophages of good quality suitable for producing virus from field isolates. Approximately, 10^{8-9} HAd$_{50}$/ml of virus were produced. The results of genome studies on DR-I virus by restriction endonuclease cleavage have been described (4). Although alveolar macrophages are readily obtained for cell culture, possible contamination with other viruses tend to restrict their use for limited applications. Search for a cell line which is fully permissive to ASFV infection should be of first priority among ASF research endeavors.

Carracosa et al. (28) reported that the virus yields in alveolar macrophages and blood monocytes, infected with wild virus, were about 1,000 HAD_{50} units/cell, a value 10-fold larger than that produced by Vero cells infected with cell-adapted virus. In this regard, we previously reported that there is about a 0.94 log differences between HAD_{50} and PFU units in expressing virus titer (9). Since the virus yields from Vero cell cultures were assayed by plaque production (28), the PFU units must be converted to HAD_{50} units before a meaningful comparison can be made. Incidentally, we have consistently obtained about 100 PFU/cell from all three types (Vero, BC, alveolar macrophages) of cells (Pan, I.C., unpublished data).

The presence of DI particles in two field isolates, Tengani and Uganda was suggested. Since the generation of DI particles can occur in cells infected with virtually any animal virus (26), its presence among ASFV isolates has been anticipated. The biological and molecular biological characterization of DI particles and the role they play in pathogenesis, persistent infections, virulence in pigs, and defense against virus invasion (27) awaits future investigation.

ACKNOWLEDGEMENTS
The following collaborators, C.J. DeBoer, W.R. Hess, W.P. Heuschele, T.S. Huang, J.E. Moulton, M. Shimizu, J. Tessler, R. Trautman, and N. Yuasa, and technical assistants, Elizabeth King, David Perkins, Brenda Rodd, James Sime, Patricia Smith, and Terry Whyard, have participated in different aspects of the work presented in this review.

REFERENCES
1. Malmquist, W.A. and Hay, D. Am. J. Vet. Res. 21: 104-108, 1960.
2. Hess, W. R. and DeTray, D. E. Bull. Epizoot. Dis. Afr. 8: 317-320, 1960.
3. Dougherty, R. M. and Harris, R. J. In: Techniques in Experimental Virology. London-New York, Academic Press Inc., Chapter 6, p. 183-186, 1964.
4. Wesley, D. and Pan, I. C. J. Gen. Virol. 63: 383-391, 1982.
5. Pan, I. C., Moulton, J. E. and Hess, W. R. Am. J. Vet. Res. 36: 379-386, 1975.
6. Pan, I. C., Huang, T. S. and Hess, W. R. J. Clin. Microbiol. 16: 650-655, 1982.
7. Reid, J. S. Am. J. Med. Technol. 39: 315-320, 1973.
8. Nakane, P. K. and Kowaoi, J. A. Cytochem. 22: 1084-1091, 1974.

9. Pan, I. C., Shimizu, M. and Hess, W. R. Am. J. Vet. Res. 39: 491-497, 1978.
10. Pan, I. C. and Hess, W. R. Am. J. Vet. Res. 45: 361-366, 1984.
11. Pan, I. C., Trautman, R., Hess, W. R., DeBoer, C. J., Tessler, J., Ordas, A., Sanchez-Botija, C. S., Ovejero, H, and Botija, Maria Carmen. Am. J. Vet. Res. 35: 784-790, 1974.
12. Pan, I. C., DeBoer, C. J. and Hess, W. R. Can. J. Comp. Med. 36: 309-316, 1972.
13. Pan, I. C., DeBoer, C. J., and Heuschele, W. P. Proc. Soc. Exp. Biol. Med. 134: 367-371, 1970.
14. Pan, I. C., Kaplan, A. M., Morter, R. L. and Freeman, M.J. Proc. Soc. Trop. Bio. Med. 129: 867-870, 1969.
15. Shimizu, M., Pan, I. C. and Hess, W. R. Am. J. Vet. Res. 37: 309-317, 1976.
16. Colgrove, G. S., Haelterman, E. O. and Coggins, L. Am. J. Vet. Res. 30: 1343-1359, 1969.
17. Boulanger, P., Bannister, G. L., Greig, A. S. Gray, D. P. and Rackerbauer, G. M. Can. J. Comp. Med. & Vet. Sci. 31: 16-23, 1967.
18. Moulton, J. E. and Coggins, L. Am. J. Vet. Res. 29: 219-232, 1968.
19. van Furth, R. and Thompson, J. Ann. Inst. Pasteur, Paris. 120: 337-355, 1971.
20. Pan, I. C. and Hess, W. R. Am. J. Vet. Res. 46: 314-320, 1985.
21. Wardley, R. C., Wilkinson, P. G. and Hamilton, F. J. Gen. Virol. 37: 425-427, 1977.
22. Wardley, R. C. and Wilkinson, P. J. Arch. Virol. 55: 327-334, 1977.
23. Casal, I., Enjuanes, L. and Vinuela, E. J. Virol. 52: 37-46, 1984.
24. Metcalf, D. Science 229: 16-22, 1985.
25. Holland, J. J. In: Field's Virology. Raven Press, N. Y., Chapter 6, p. 77-99, 1985.
26. Huang, A. S. and Baltimore, D. In: Comprehensive Virology (eds. Franke Conrat, H. and Wagner, R.R.) p. 73-116. Plenum, New York. 1977.
27. Huang, A. S. and Baltimore, D. Nature 226: 325-327, 1970.
28. Carrascosa, A. L., Santaren, J. F. and Vinuela, E. J. Virol. Methods 3:303-310, 1982.
29. Moore, D. J. J. S. Afr. Vet. Ass. 49: 259-264, 1979.
30. Mauer, F. D., Griesemer, R. A. and Jones, T. C. Am. J. Vet. Res. 19: 517-539, 1958.
31. Enjuanes, L., Cubero, I. and Vinuela, E. J. Gen. Virol. 34: 455-463, 1977.
32. Moura Nunes, J. F. and Nunes-Petisca. Proc. CEC/FAO Res. Seminar held in Sassari, Sardina. EUR 8466 EN 132-142, 1983.
33. Wardley, R. C. Immunology. 46: 215-220, 1982.
34. Shimizu, M., Pan, I. C. and Hess, W. R. Am. J. Vet. Res. 38: 27-31, 1977.
35. Hess, W. R. and Pan, I. C. FAO/EEC Experts Consultation on Classical Swine Fever and African Swine Fever. Monograph, 602-611. Hannover, W. Germany, 1979.
36. Sanchez-Viscaino, J. M., Slauson, D. O., Ruiz-Gonzalvo, F. and Valero, F. Am. J. Vet. Res. 42: 1335-1341, 1981.
37. Cheville, N. F. and Mengeling, W. L. Lab. Invest. 20: 261-274, 1969.
38. Winkler, G. C. and Cheville, N. J. Leuk. Biol. 38: 471-480, 1985.
39. Moulton, J. E., Pan, I. C., Hess, W. R., DeBoer, C. G. and Tessler, J. Am. J. Vet. Res. 36: 27-32, 1975.

40. Pan, I. C., Shimizu, M. and Hess, W. R. Am. J. Vet. Res. 41: 1357-1367, 1980.
41. Breese, S. S. Jr., and DeBoer, C. J. Virology 28: 420-428, 1966.
42. Compagnucci, M., Martone, F. and Vaccaro, A. Vet. Italiana. 20: 395-402, 1969.
43. Pan, I. C., Wang, C. T., Yeh, Y, C., Pan, I. J. and Chen, H. C. Bull. Inst. Zool., Academia Sinica. 1: 73-87, 1962.
44. Pan, I. C., Whyard, T. C. and Hess, W. R. Fed. Proc. No. 3051, p. 1939, 1984.
45. Grieg, A. and Plowright, W. J. Hyg. (Camb) 68: 673-682, 1970.
46. McVicar, J. W. Am. J. Vet. Res. 45: 1535-1541, 1984.
47. Schlafer, D. H. and Mebus, C. A. Am. J. Vet. Res. 45: 1353-1360, 1984.

10

APPROACHES TO VACCINATION

U. KIHM, M. ACKERMANN, H. MUELLER and R. POOL
Federal Vaccine Institute, 4025 Basle, Switzerland

ABSTRACT

Pigs surviving an African swine fever virus (ASFV) infection are protected against reinfection with the homologous virus strain. However, experiments with live or inactivated vaccines have not led to sufficient protection (1-5). Here we show data using inactivated, detergent-treated preparations originating from pigs infected with a highly virulent ASFV, rendering a fairly good survival rate (50-100 %) of experimentally infected pigs. Also, virulent virus propagated in pig leucocyte cultures followed by inactivation conferred protection while 80-100% of control pigs died after infection. The use of complete Freund's adjuvant improved the survival rate considerably. However, none of the preparations used as vaccine was able to prevent ASFV infection and replication. Consequently, the vaccinated pigs and thereafter experimentally infected proved to be virus carriers at least for 91 days. Virus persistence was demonstrated in vivo by infection of healthy pigs with blood sampled from carriers and in vitro by inoculation of pig leucocyte cultures. However, spontaneous virus transmission from surviving pigs to untreated in-contact pigs was not observed. Our serological results (ELISA, IIF, IEOP) indicate that protection is not based on antibodies to VP 73 but on factors preserved in inactivated antigens prepared from virulent virus.

INTRODUCTION

Research into African swine fever (ASF) has still not led to an effective vaccination for the protection of pigs. In the past, attenuated live virus vaccines have been investigated (1, 2). Pigs immunized with live virus vaccines were able to survive challenge with homologous virulent ASF virus (ASFV). Following challenge, however, they proved to be virus carriers. Furthermore, field experiments with live attenuated preparations have resulted in the development of chronic infections and death in vaccinated animals from which virus was isolated (1).

Little success has been achieved in the immunization of pigs with inactivated antigens for preventing infection (3, 4, 5). Although a serological immune response was observed in some cases with the detection of various antibodies, the immunized pigs were not protected.

In contrast, a spleen suspension from an ASF-infected pig, treated with a nonionic detergent and mixed with complete Freund's adjuvant and twice inoculated into pigs conferred protection against death after oral inoculation of homologous ASFV (6). Seven out of ten vaccinated pigs survived the virus infection, but all animals showed characteristic ASF signs. All surviving animals recovered completely, but they remained persistently infected. Two months after the infection with homologous virus, the seven surviving pigs were challenged with a heterologous virus. Chronic infection became established in all seven pigs. Five animals died between 9 and 18 days after infection.

In order to elucidate the protection—conferring factors of the vaccine mentioned above three experiments were performed.

Specifically, the effects of adjuvants, multivaccination and antigens prepared from different sources were determined.

MATERIAL AND METHODS

Virus

The ASF virus used for all the vaccination and challenge studies was the Perpignan strain isolated in France in 1964. The isolate belongs to the group of highly virulent ASF viruses and has been passaged twice in pigs. The spleen of the second virus passage produced in 1968 was stored at $-70^{\circ}C$ and used for antigen preparation and challenge (spleen 68). Virus infectivity was determined by the hemadsorption test (HAD) in pig leucocyte cultures according to the method of Malmquist (7).

Virusisolation

Samples obtained for virus detection from serum, erythrocytes, buffy coat and from 10% suspension of tonsils, spleen and lymph nodes were inoculated into pig leucocyte cultures. The cultures were examined for development of cytopathic effect (CPE) and HAD. In experiment 2, heparinized blood from persistently infected pigs was inoculated into seronegative pigs (2,5 ml intramusculary (im) and 2,5 ml orally).

Antigen preparations

Preparation A

The spleen 68 originated from a pig that had died 6 days post-infection (pi) with the virulent Perpignan strain. The spleen was homogenized in Minimal Essential Medium (MEM) Eagle to make a 10 % suspension. The suspension was then clarified by centrifugation at 6000 rpm for 20 minutes (min) to yield a supernatant fluid containing $10^{6.7}$ hemadsorption units per ml (HAD_{50}/ml).

Preparation B

The spleen suspension was prepared as indicated for preparation A. The spleen (spleen 83) originated from a pig that had died 11 days after inoculation of spleen 68. The virus content was $10^{7.7}$ HAD_{50}/ml.

Preparation C

Erythrocytes were obtained from a pig that had died 5 days after inoculation of spleen 68. The erythrocytes were washed twice with physiological saline solution and then frozen at -70°C until use. After thawing, the erythrocytes were suspended in physiological saline. The virus content was $10^{6.4}$ HAD_{50}/ml.

Preparation D

Infected pig leucocyte cultures were harvested at 5 days pi with spleen 68 by freezing and thawing. The virus content was $10^{7.7}$ HAD_{50}/ml.

Placebo

A 10 % spleen suspension was prepared as outlined in preparation A from a noninfected pig.

Inactivants

The virus suspension were treated with the nonionic detergent n-octyl beta-D-glucopyranoside (OBG[1]). For complete inactivation a 1% solution was used in the case of preparation A, B and D. Preparation C was treated with a 2% solution of OBG. In one experiment the spleen 68 was first treated for two hours with ß-propiolactone 0,1 % (BPL[2])

---

[1] Sigma, St. Louis, USA

[2] Fluka AG, Buchs, Switzerland

followed by 1% OBG (Preparation A*). In all cases absence of residual live virus was determined in pig leucocyte cultures by the HAD test.

Adjuvants

Complete Freund's adjuvant and an oil adjuvant kindly supplied by Rhone Mérieux, F-Lyon were used. The antigens were mixed 1:1 with the adjuvants. The oil consisted mainly of paraffin (Marcol) and Arlacel.

Pigs

Landrace pigs, each weighing approximately 20 kg were vaccinated im in the neck. Blood samples were collected from all pigs before vaccination, before challenge and at various intervals after challenge. For infection one ml of diluted spleen 68 virus was given orally ($10^{4.7}$ HAD_{50}).

Vaccination

The experimental design was always based on at least two vaccinations followed by oral inoculation of homologous ASFV. Three experiments were carried out.

In the first, the effect of preparation A mixed with two different adjuvants was evaluated. At the same time the specificity of the protection conferred by the prepared antigens was determined in comparison to placebo. The vaccines used in this experiment are indicated in Table 2.

In the second experiment the effect of multivaccination with adjuvanted and nonadjuvanted preparation A was determined. The specifications of this experiment are given in Table 3. Thirteen pigs surviving this experiment were tested for virus persistence.

In the third experiment, the effect of five different antigen preparations was tested. The antigens used are listed in Table 6. Serum samples were collected at intervals for antibody determination using the following methods:

Indirect Immunofluorescence (IIF)

The method used is described by Bool et al (8). Briefly, CV_1 monolayer cells infected with a cell adapted ASFV (Lisbon 60) were fixed with acetone. The slides were overlayed with test serum and then stained with anti-pig serum FITC.

ELISA

Assays were carried out using the procedures of Bommeli et al (9) and Sanchez Vizcaino et al (10). The VP 73 antigen was kindly supplied by Dr. M. Lombard, Rhône Mérieux, F-Lyon).

Immune electroosmophoresis (IE OP)

The tests were conducted according to the method of Pan et al (11).

RESULTS

Clinical disease

Both vaccinated and nonvaccinated pigs in all experiments showed fever (up to 42°C), anorexia and apathy 3-5 days after challenge. Some showed cyanotic patches on the ears, tails or legs. Constipation or bloody diarrhea were also observed in some pigs. A correlation between nonprotected pigs and the occurrence of certain clinical signs could not generally be determined. However, the changes in blood coagulation observed during fever seemed to increase the likelihood of a fatal outcome. The blood of some viraemic pigs coagulated to a gel immediately after collection and the preparation of serum was no longer possible. All 13 control animals with coagulation defects died but only 24 of 42 vaccinated animals with the same defect succumbed (Table 1). Unprotected pigs died between day 8 and day 16 pi. The surviving pigs generally recovered within 21 days and looked healthy with a normal growth rate until slaughter.

Table 1. Correlation between coagulation changes and mortality (%) in vaccinated and nonvaccinated pigs.

coagulation changes		no coagulation changes	
controls	vaccinated	controls	vaccinated
100	57	66	39

First experiment (Table 2)

In order to determine the effect of adjuvanted preparation A and of placebos, 3 groups, each consisting of 8 pigs, were vaccinated. In the group vaccinated with preparation A and Freund's adjuvant 4 animals survived. The preparation A containing oil adjuvant, however, only protected 2 pigs against death. In the control groups vaccinated with various preparations lacking ASF antigen, a single pig survived.

Table 2. Results of challenge following two inoculations using OBG–
inactivated ASFV with Freund's adjuvant or oil adjuvant versus placebo.

Inoculum[a]	Adjuvant	Number of pigs surviving challenge[b]
Preparation A whole spleen	Freund	4/8[c]
Preparation A whole spleen	Oil emulsion	2/8
Controls without ASF antigen:		
. whole spleen + OBG + Freund		0/2
. whole spleen + OBG + oil emulsion		0/2
. --- OBG + Freund		0/2
. --- OBG + oil emulsion		1/2
. no vaccination		0/2

[a] 2 ml given im twice.
[b] challenge of immunity by oral inoculation of homologous ASFV ($10^{4.7}$ HAD_{50}/ml) two weeks after the last vaccination.
[c] survivors per number of challenged pigs.

Unprotected vaccinated and control pigs died between days 7 and 16 pi. During the post-infection period virus was isolated from the serum of all pigs at least once. In both vaccinated groups the clinical signs of the survivors disappeared at day 17 pi. The surviving pigs recovered completely and could not be distinguished from noninfected pigs.

Second experiment (Table 3)

The protection rate of the pigs was not improved with the multivac-cination schedule. Only 2 of the 6 pigs vaccinated 8 times survived the infection. On the other hand two vaccinations with preparation A and Freund's adjuvant led to a high rate of survival (5/6) and even two injections without adjuvant conferred protection on 3 pigs (3/6). Controls and vaccinated but unprotected pigs died between days 7 and 19 pi.

In the 23 days following challenge, virus isolation from the serum of all surviving animals was performed 8 times at regular intervals, the results of which are compiled in Table 4. The high percentage of protected animals in group III was reflected by the short viraemic phase compared to the other vaccinated or control pigs.

Table 3. Results of challenge following two or eight inoculations using
OBG-inactivated ASFV with or without Freund's adjuvant.

Group	Inoculations of Preparation A whole spleen	Adjuvant	Number of pigs surviving challenge
I	8^a	Freund	$2/6^c$
II	8	----	$2/6^c$
III	2^b	Freund	$^e5/6^d$
IV	2	----	$3/6^d$
V	no vaccination	----	$1/6$

[a] 2 ml given im 3 to 4 days apart or
[b] 2 weeks apart.
[c] challenge of immunity by oral inoculation of homologous ASFV $10^{4.7}$
HAD_{50}/ml) 3 days or
[d] two weeks after the last vaccination.
[e] one pig died during blood sampling on day 16 pi.

Table 4. Virus isolation from sera of vaccinated and nonvaccinated
pigs after challenge with virulent ASFV. Isolation period: day 2 until
day 23 pi.

Group	Number of virus isolation assays[a]	Positive virus isolation assays
I	30	14 (47 %)
II	39	18 (46)
III	46	9 (20)
IV	37	18 (49)
V controls	34	16 (47)

[a] the number of assays is different in each group due to the different
number of survivors at the time.

At day 30 pi 4 out of 13 healthy surviving pigs were slaughtered
(No. 16, 17, 18 from group III and No. 24 from group IV) and an organ
pool (tonsils, spleen, lymphnodes) of each pig was made. Virus was
detected from organs of pig No. 16, 17 and 24 by HAD. In pig No. 17
virus was also found in the bone marrow.

The other 9 surviving pigs from all groups were used for further

Table 5. Virus isolation from the blood of ASF-infected pigs during 107 days pi. Samples of leucocytes and erythrocytes were inoculated into pig leucocyte cultures and twice blind passaged.

Group	No.pigs	days post infection																				
		2	4	7	9	11	14	18	23	49	58	63	70	73	77	80	84	87	91	94	101	107
I	2	-	-	-	+	-	-	-	-	-[a]	-[a]	-	-	-	-	-	+	+	-	-	-	-
	4	-	+	+	+	+	-	-	-	-	-	-	-	-	-	-	-	-	-	-	-	-
II	8	-	+	+	+	+	-	-	-	-[a]	-	-	-	-	-	-	-	-	-[b]	-	-	-
	11	-	-	-	+	-	+	+	-	+	-	-	-	-	+	-	-	-	+	-	-	-
III	14	-	-	+	+	-	-	-	-	-	-	-	-	-	-	-	-	-	+	-	-	-
	15	-	-	+	+	-	-	-	-	-	-	-	-	-	+	+	-	-	-	-	-	-
IV	21	-	-	+	+	+	-	-	-	-	+	+	-	-	-	-	-	-	-	-	-	-
	23	-	-	+	+	+	-	-	-	-	+	-	-	-	-	-	-	-	-	-	-	-
V	27	-	-	-	+	-	-	-	-	-[a]	+	-	-	-	-	-	-	-	-	-	-	-
contact pigs[c]	31									-	-	-	-	-	-	-	-	-	-	-	-	-
	32									-	-	-	-	-	-	-	-	-	-	-	-	-

[a] ASFV transmitted by infected blood to another pig kept in isolation.
[b] slaughter due to intercurrent disease (tail abscess).
[c] kept in contact with pigs No. 2, 8, 14, and 21 from day 49 pi onwards.
+ = virus detected in pig leucocyte culture.
- = no virus detected in pig leucocyte culture.

virus isolation studies. Table 5 shows a summary of virus isolation tests in these pigs during a period of 107 days pi. Virus was detected regularly during 14 days pi and also later (HAD). In contrast, 2 contact animals introduced from day 49 pi onwards were not infected and remained healthy and seronegative. On the other hand pigs kept in isolation were infected by blood transmission given im and orally, although in certain cases no virus was detected by HAD.

All 9 pigs except one (No. 11) remained healthy during the observation period and showed a normal growth rate. At day 107 pi all the pigs were sacrificed, and hemadsorption tests carried out to demonstrate virus in blood, tonsils, spleen and lymph nodes. The results proved to be negative.

Third experiment (Table 6)

The different vaccine preparations elicited different percentages of protected pigs. The fairly high protection rate, which had already been established with the whole spleen suspension (preparation A), was confirmed. The use of BPL with additional OBG treatment had little, if any influence on the survival rate (preparation A*). In contrast, the whole spleen suspension (preparation B) from a pig which had died at day 11 pi, containing ten times more virus, was found to be less effective with regard to survival after challenge. The vaccine containing washed erythrocytes (preparation C) of a pig infected with spleen 68 protected the same number of pigs as preparation A. The erythrocytes were harvested from a pig which had died on day 5 pi. The virus titer was similar to that of the spleen 68 used for preparation A.

However, best results were achieved with an ASF antigen produced in leucocyte cultures (preparation D). Here all 5 animals were protected.

None of the vaccine preparations prevented virus infection and replication after challenge. Nonetheless, the number of surviving animals differed appreciably from one group to another (Table 6).

Between days 4 and 15 pi virus was isolated at least twice from the sera of all animals regardless of their immune status. The surviving animals were not able to eliminate the virus. Although ASF virus was not isolated from the serum 22 days pi, the spleen, tonsils and different regional lymphnodes of the protected pigs were still infected (Table 7).

Table 6. Challenge results of pigs vaccinated with OBG inactivated ASFV originating from infected leucocyte cultures, erythrocytes and whole spleens of infected pigs.

Inoculum [a]	Virustiter before inactivation (HAD_{50}/ml)	Number of pigs surviving challenge [g]
Preparation A whole spleen [b]	$10^{6.7}$	4/6
Preparation A* whole spleen [c]	$10^{6.7}$	3/6
Preparation B whole spleen [d]	$10^{7.7}$	1/6
Preparation C erythrocytes [e]	$10^{6.4}$	4/6
Preparation D pig leucocyte cultures [f]	$10^{7.7}$	5/5
Controls	---	1/6

[a] two inoculations: 2 weeks apart. All inocula contained Freund's adjuvant.
[b] harvested from a pig which died 6 days pi.
[c] same as [b]. Treated with 0.1% BPL for 2 hrs before OBG.
[d] harvested from a pig which died 11 days pi.
[e] harvested from a pig which died 5 days pi.
[f] the infected cultures were harvested 5 days pi.
[g] challenge of immunity by oral inoculation of homologous ASFV ($10^{4.7}$ HAD_{50}/ml) two weeks after the last vaccination.

Table 7. Isolation of virus from serum, spleen, tonsils and lymphnodes of vaccinated pigs in Experiment 3.

Isolation of virus from pigs	Serum	Spleen	Tonsils	Lymphnodes
died between day 8-15 pi	12/12 100%	11/12 92	10/12 83	11/12 92
survived [a] until day 22 pi	0/17 0%	7/17 40	7/17 40	5/17 30

[a] All survivors were sacrificed at day 22.

Serology

Fig. 1 shows the ELISA antibody titers of the pigs vaccinated with

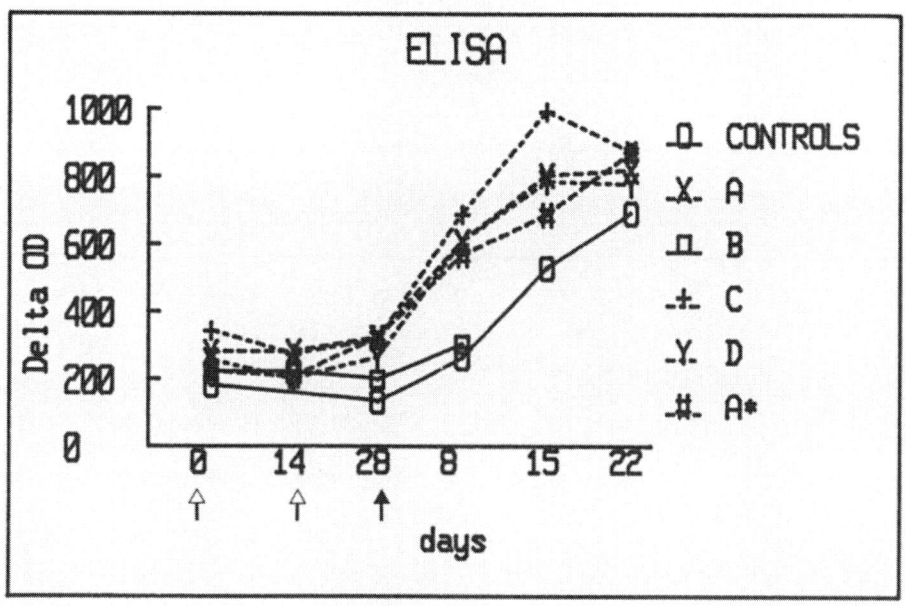

Fig. 1. Evaluation of antibodies to VP 73 following vaccination and ASFV infection. Antigens used in the various vaccines are described in experiment 3. Microtiterplates were coated overnight (4°C) with 120 μl VP 73 per well (positive antigen). The control antigen consisted of a cytoplasmic extract from noninfected Vero cells. After 3 washings pig sera diluted 1:40 were added to 4 wells (2 sensitized with positive and 2 with control antigen). After incubation with anti-pig IgG–peroxidase the wells were filled with substrate (ABTS) solution (9). The optical density (OD_{405}) given on the ordinate was calculated by forming the difference between OD of the sera with positive and control antigen (Delta OD). Median values of Delta OD have been plotted. Open arrows indicate days Of vaccination. The dark arrow points to the day of infection with ASFV. Note: - in group B all pigs except one died before day 15 pi.
- days 0, 14, 28 represent days after the first vaccination.
- day 8, 15, 22 represent days pi.

different antigen preparations (Experiment 3) and after infection with ASFV. There were no significant antibody titers to VP 73 after two vaccinations. Most pigs showed an increase in antibody titer 8 days after infection except for those in group B and the control animals. In each of these two groups 5 out of 6 pigs died. All animals in the other groups showed an antibody reaction, irrespective of whether they survived or died.

Immunofluorescent antibody titers are presented in Table 8. These

Table 8. Immunofluorescent antibody titres following vaccination and experimental infection with ASFV. Median antibody concentrations were evaluated by indirect immunofluorescent technique.

Blood sampling	Vac [a]	Vac [b]	Inf [c]	4 [d]	8 [d]	11 [d]	15 [d]	22 [d]
Group [e) A$_*$	neg [f]	neg	5 [g]	5	80	320	960	1280
A	neg	neg	5	5	80	320	960 [h]	1280 [h]
B	neg	neg	5	5	30	160	ND [h]	ND [h]
C	neg	5	neg	5	60	320	960	1280
D	neg	neg	5	5	80	320	640	1280
Controls	ND	ND	ND	neg	neg	10	80	320

[a] first vaccination 28 days before infection.
[b] second vaccination 14 days before infection.
[c] day of oral infection with virulent ASFV.
[d] days pi.
[e] groups refer to the respective antigen preparations used as vaccines.
[f] neg = no fluorescence detected with undiluted serum.
[g] reciprocal of serum dilution still revealing fluorescent foci.
[h] all pigs except one died before day 15 pi.
ND = not done.

results were comparable to the ELISA values. The antibody titers determined before infection were low and increased markedly from day 4 to day 11 pi. However, although some pigs showed a positive antibody response after vaccination, this did not necessarily indicate protection (Table 9). Moreover, even an immediate increase in antibody titer following infection of the pigs vaccinated with preparations A, A* C and D failed to correlate with protection. All but one of the nonprotected vaccinates had high antibody titers.

In addition, no IEOP antibodies were demonstrated after vaccination.

Table 9. Correlation between the presence of IIF-antibodies and mortality in vaccinated pigs.

Immune status before challenge	No. of pigs	Mortality
seropositive	22	9/22 = 41%
seronegative	7	3/7 = 43%

However, as early as 4 days pi some pigs already showed evidence of IEOP antibodies (Table 10). At day 8 pi antibodies were found in most of the pigs in groups A, A*, C and D. By contrast, only three pigs from group B and only one of the controls revealed antibodies. Thus, failure to develop early antibodies (8 days pi) correlated with poor protection and the survival in these two groups was low. On day 11 pi all but one of the vaccinated pigs were seropositive and protection was no longer predictable.

Table 10. IEOP antibodies detected in vaccinated and control pigs after infection with ASFV.

	days pi[a]	4	8	11
Groups[b] A*		$4^c/6^d$	5/6	5/6
A		2/6	4/5	5/5
B		0/6	3/6	5/5
C		0/6	4/6	6/6
D		0/6	4/5	5/5
Controls		0/6	1/6	4/6

[a] at the day of infection no IEOP antibodies could be detected.
[b] groups refer to the respectiv antigen preparations used as vaccines.
[c] number of seropositive pigs.
[d] number of pigs tested.

DISCUSSION

In spite of the assumption that immunization with inactivated ASF antigens was likely to prove difficult (3, 4), earlier experiments had shown that pigs could be protected from death by the administration of an inactivated whole spleen vaccine (6). The presence of a maximum possible number of ASF-specific antigens could be assumed in spleens of infected pigs. The spleens were obtained from pigs which had died after infection with a virulent ASFV (Perpignan strain). These preparations contained structural as well as nonstructural viral and host protein which could serve a protective function (6). For these reasons, as many native "antigens" from homologous tissues as possible had to be administered to the pigs in a preparation free from live ASFV. A method rendering both, reliable inactivation and conservation of immunogenic properties was therefore necessary. Tests with detergents have led to the use of OBG, which provided complete inactivation of the ASFV. The lack of a convincing method to detect neutralizing antibodies (12) emphasized the need to stimulate an immune response as broad as possible. Therefore we employed complete Freund's adjuvant. Being aware of all the disadvantages of such an adjuvant (e.g. tissue necrosis, no applicability in the field), we considered them to be of no relevance in the case of experimental vaccines.

Innocuity

In various preliminary studies using inactivated preparations, we never detected residual live virus in pig leucocyte cultures. All these preparations proved to be non-infectious in the pig. Even when the dose was increased 200-fold no evidence of residual live ASFV was observed (data not shown). By contrast all the pigs in a group inoculated with insufficiently inactivated spleen suspension became ill and died (data not shown).

A transient increase in body temperature after vaccination was noted in some cases due to complete Freund's adjuvant. Local tissue necrosis at the site of injection behind the ear remained within normal limits when the animals were immunized twice with an interval of 14 days between each injection. In Experiment 2, however, repeated administration every 3-4 days led to relatively marked tissue reactions. By contrast, injection of the spleen suspension without adjuvant produced no reaction.

Clinical disease

Following experimental infection no clinical differences between vaccinated and non-vaccinated pigs were observed with the exception of the survival rate. Almost all animals showed fever, loss of appetite, and apathy; at times, cyanotic patches on the ears, tail, legs, and abdomen were also observed, as were constipation or diarrhoea, and occasionally epistaxis, ataxia, and anaemia. All these reactions were distributed unevenly among the various vaccinated groups and controls. Likewise no obvious pattern in the duration of the disease either to recovery or to death could be observed. Sometimes it seemed that the onset of disease was earlier and the course more severe in the vaccinated animals than in the controls. A similar phenomenon has been reported after sensitization with ASFV (13).

The changes in blood coagulation gave an indication (Table 1) as to whether an animal was likely to survive the infection or not. In the vaccinated groups 50 % of the animals with disturbances in blood coagulation survived. In contrast, all the controls with coagulation problems died. The mortality rate among the vaccinated pigs, however, increased to 80 % when the coagulation changes lasted longer than three days. A correlation would thus appear to exist between mortality and changes in blood coagulation.

Whether or not the same blood coagulation factors were affected as those described by Edwards (14) was not investigated in our experiments.

Protection of pigs using inativated ASF vaccines

Using the spleen antigen mixed with complete Freund's adjuvant in all our experiments we were able to protect over 50% of the pigs from death following homologous ASFV infection. The results of these investigations confirmed our earlier results (6) but are in contrast to those of other authors (4, 5). Since our experimental design differed from that used by these authors in respect of the strain of virus used, the inactivation and the mode of infection, our experiments led therefore to different results. The use of a suitable adjuvant seemed to be crucial. Antigens adjuvanted with oil conferred less protection than those with complete Freund's adjuvant (Table 2). It would seem that tubercle bacilli are important for the induction of a protective immunity. Furthermore repeated antigen administration, both with and without adjuvant, failed to enhance the protective effect, although it is possible that a suboptimal vaccination schedule was selected (Table 3).

However, ASF antigens obtained both from erythrocytes of an infected pig and from infected pig leucocyte cultures proved to be potent protective immunogens (Table 6). Since the erythrocytes obtained from a viraemic pig were washed twice before virus inactivation, it can be concluded that the protection-inducing factors are related with the virus-erythrocyte complex. Furthermore these factors seemed to originate from cells competent for virus replication since pigs treated with a vaccine consisting of ASFV-infected leucocyte cultures displayed an excellent degree of protection. It was not possible to determine whether protection was conferred by a specific virus population alone, or whether additional protective factors were produced which might bind as well to erythrocyte membranes. However, the protective effect was specifically virus-dependent, various control preparations having failed to induce any protection (Table 2).

Interestingly, the concentration of virus in the erythrocyte preparation ($10^{6.4}$ HAD_{50}) was 2 to 20 times lower than that in all other preparations, and yet still the rate of protection was very good (Table 6). On the other hand, no protection could be induced with preparation B despite of the high virus concentration ($10^{7.7}$ HAD_{50}). There are at least two obvious explanations for this phenomenon:

First, the harvest time and the number of passages of the virus used for the vaccine preparation may be crucial. The spleen 68 used in preparation A was harvested 6 days pi and the spleen 83 for preparation B at the 11th day pi. Furthermore, spleen 83 represented an other pig passage of spleen 68.

Second, different virus populations arising in the early or late phase of viraemia have to be considered.

As far as we are aware, no one has yet achieved a similar rate of protection using inactivated antigens. It therefore has to be assumed that, in the case of the detergent OBG, an agent was found which was able to destroy the infectious virus but still to maintain the antigens responsible for protection.

Although the protection rates achieved in the various experiments are highly encouraging, it has to be pointed out that infection with and replication of ASFV could not be prevented. All the pigs were thus found to be viraemic between the 4th and 18th day following infection. The virus subsequently persisted, in spite of high antibody titers, until at least the 91st day pi. At irregular intervals, ASFV was demonstrated in erythrocytes or leucocytes of the persistently infected, but healthy, pigs by inoculating leucocyte cultures. Natural infection of contact pigs, however, was not observed during this period (Table 5). When blood of such animals was inoculated into seronegative pigs, these died as a result of acute ASFV infection. Since blood-sucking ticks of the genus ornithodorus do act as transmission vectors, a recurrent viraemia in pigs may be an important factor for the dissemination of the disease.

Serology

The antigens for our serological studies were prepared from a cell-adapted heterologous ASFV strain (Lisbon 60). Most probably the antibodies detected by ELISA, IIF and IEOP respectively were directed largely against VP 73. In our experiments no serovoncersion has been shown after two or eight vaccinations. Following challenge, however, a booster effect was observed in vaccinated animals. Eight days pi high antibody titers were detected in animals vaccinated with preparations conferring protection. In contrast controls and vaccinated pigs with low survival rates (group B) showed a delayed antibody response. Moreover individuals with an early antibody increase were not necessarily protected (Table 9). From these observations we many conclude the following:

I)   A priming effect in vaccinated pigs could be detected.

II)   We are not able to show the immune response relevant for preventing death.

III) Antibodies to VP 73 seem not to be specific concerning protection. Nevertheless, they may be useful in seroepidemiological studies, since VP 73 is specified by a large variety of ASFV strains.

IV)   Further experiments have to demonstrate antibodies to protection-inducing antigens. In addition the role of cell-mediated immunity in our system has to be determined.

REFERENCES

1. Manso Ribeiro, J., Petisca, J.L.N., Frazao, F.L. and Sobral, M. Bull. Off. Int. Epiz. 60: 921-937, 1963.
2. Sanchez Botija, C. Zooprofilassi 18: 578-607, 1963.
3. De Tray, D.E. In: Advances of Veterinary Science Vol. 8 (Eds. C.A. Brandly and E.L. Jungherr), Academic Press, New York, 1963, pp. 299-333.

4. Stone, S.S. and Hess, W.R. Am. J. Vet. Res. 28: 475–481, 1967.
5. Forman, A.J., Wardley, R.C. and Wilkinson, P.J. Arch. Virol. 74: 91–100, 1982.
6. Bommeli, W., Kihm, U. and Ehrensperger, F. In: Proceedings of a CEC/FAO Research Seminar, Sassari, 1981 (Ed. P.J. Wilkinson), Published by the Commission of the European Communities, 1983, pp. 217-223.
7. Malmquist, W.A. and Hay, D. Am. J. Vet. Res. 21: 104–108, 1960.
8. Bool, P.H., Ordas, A., Sanchez-Botija, C.S. Bull. Off. Int. Epiz. 72: 819–839, 1969.
9. Bommeli, W., Kihm, U., Lazarowicz, M. and Steck, F. In: Proceedings of the Second International Symposium of Veterinary Laboratory Diagnosticians, Lucerne, 1980, pp. 235-239.
10. Sanchez Vizcaino, J.M., Crowther, J.R., Wardley, R.C. In: Proceedings of a CEC/FAO Research Seminar, Sassari, 1981 (Ed. P.J. Wilkinson), Published by the Commission of the European Communities, 1983, pp. 297-325.
11. Pan, J.C., De Boer, C.J. and Hess, W.R. Can. J. Comp. Med. 36: 309-316, 1972.
12. Hess, W.R. Adv. Vet. Sci, Comp. Med. 25: 30–69, 1981.
13. Edwards, J.F., Dodds, W.J. and Slauson, D.O. Am. J. Vet. Res. 46: 2058-2063, 1985.
14. Edwards, J.F., Dodds, W.J. and Slauson, D.O. Am. J. Vet. Res.: 45: 2414-2420, 1984.

11
EPIDEMIOLOGY OF AFRICAN SWINE FEVER VIRUS

Y. BECKER

Department of Molecular Virology, Faculty of Medicine, The Hebrew
University, 91 010 Jerusalem, Israel

ABSTRACT
    African swine fever virus (ASFV), which prevails in Eastern and
Southern Africa as a mild disease-causing agent in warthogs, found a most
sensitive host in the domesticated pig. In the latter, ASFV affects the
reticuloendothelial cells and causes the death of the host without a
response from the immune system. ASFV made its appearance in Europe in
the 1960s, infecting pigs in the Iberian peninsula, and causing great
economic losses as a result. ASFV spread to France, causing local
epidemics in 1964, 1968 and 1974. The year 1978 marked the appearance
of ASFV in Malta and Sardinia, as well as in Brazil, followed by epidemics
in the Dominican Republic and Haiti (1979) and Cuba (1980). The ASF
epidemics were halted by destruction of the pig populations in the
affected countries. This led to eradication of the disease, except in
Sardinia, where the virus escaped into the wild pig population. After a
hiatus of several years, ASFV caused an epidemic in Belgium in 1985 and
in The Netherlands in 1986. Rapid methods of diagnosis, strict control of
pig movements, and the elimination of infected pigs put a stop to the
outbreak.

The spread of African swine fever (ASF) in Africa followed the
distribution of warthogs on that continent, but the major difficulty in
understanding the epizootiology has been accounting for the transmission
of the virus from the warthog population to domestic swine (1, 2).
Plowright (2) demonstrated that the argasid tick *Ornithodorus moubata*
is an extremely effective reservoir and vector of ASFV in some areas of

Dedicated to the memory of Dr. J. Manso Ribeiro, Lisbon, Portugal.

Eastern and Southern Africa. These ticks, carried by warthogs, are transferred to territory used for domestic pigs, thus passing on the virus. In addition, ASFV can infect domestic pigs when the latter are fed on warthog carcasses (2).

The development of extensive commercial links and air traffic between various countries made possible the transfer of ASFV from its endemic areas in Africa to other continents. The ability of ASFV to remain in pig meat products even after processing makes the remains of food from boats or airplanes that are fed to pigs at locations near harbors and airports a possible source of infection. The history of ASF outbreaks, as seen in Table 1, is a sad reminder of the extent of ASFV transmission and the enormous economic damage caused at the receiving end. In the most recent outbreaks, the source of ASFV infection was traced to swill fed to pigs in the vicinity of harbors and airports. The availability of improved diagnostic tools to shorten the time between the start of an outbreak and

Table 1. Spread of ASFV

Country	Year	Ref.	First outbreaks
Africa (East and South)	Described 1921	1,2	
Iberian Peninsula	1960	3,4	
France	1964,1968,1974	5	
Malta	1978	6	Piggeries near port and airport
Sardinia	1978	7	
Cuba	1971,1980	8	
Brazil (South)	1978	9,10	
Dominican Republic & Haiti	1979	11	
Belgium	1985		
The Netherlands	1986		C. Terpstra, personal communication

ASFV appeared in Spain in 1960 in four epizootic waves (12) with maximum peaks in the years 1963 (1250 farms affected), 1967 (3233 farms), 1971 (1714 farms) and 1977 (1894 farms). With the introduction of a national eradication program, the number of affected farms gradually decreased: 1978 - 1428 farms, 1979 - 1044 farms, and 1980 - 447 farms. In 1977, 318,292 pigs were sacrificed and farmers were paid 762.2 x 10^6 pesetas (US$7,622,000). In subsequent years, the number of sacrificed pigs was 245,100 (1978), 245,000 (1979), and 110,000 (1980), with total compensation for the three years amounting to US$13,480,000 (12). Botiza (13) demonstrated in 1963 that ASFV is present in Spain in the tick *Ornithodorus erraticus*; hence the prevalence of the virus in that country. Improvements in zoosanitary policy, better control of slaughter houses, and reorganization of the pig industry, coupled with an extensive eradication program, are some of the factors which led to a marked reduction in the number of infected pig farms. In Portugal (14), peaks of ASF appear at intervals of four to five years. After 1977, a progressive decrease in the number of infected pigs was observed. Yet, ASF remains endemic in Portugal.

Outbreaks in France (Table 1, ref. 5) were limited to the Southern part of the country. In some instances, ASFV was carried by pigs illegally transferred across the border from Spain.

The year 1978 marked a severe spread of ASFV to the Republic of Malta, and Sardinia, islands in the Mediterranean Sea, and also to Brazil (Table 1). The virus appeared in Malta at the beginning of March 1978, with outbreaks near the port of Valletta and the airport of Luga, where pigs were fed with waste from ships and/or airplanes. Within 45 days, 600 out of 1600 farms were involved. To stop ASFV from spreading from Malta, the Commission of European Communities (CEC) undertook to donate

5 million units of account to the government of the Republic of Malta to aid in an eradication program of the entire pig population, so as to ensure elimination of ASF (6).

The situation in Sardinia (7) was, at first, similar to that in Malta. The first outbreak occurred on March 17, 1978, in Cagliari province in a small group of domestic pigs and in a herd of free-ranging pigs nearby. The source of infection is believed to have been swill from the port or airport of Cagliari that was used to feed the pigs. After the original outbreak, despite the application of control measures, the infection spread to 14 other districts in the same province. In May 1978, the disease appeared in another province, and it continued to spread in 1980. The virus was found in wild boars, indicating its spread into the free-living wild pig population on the island. Eradication measures were used, and 42,500 pigs were slaughtered, with compensation being given to the farmers.

ASFV arrived in Brazil in 1978 through garbage feeding on a low technology hog farm (9, 10). In an emergency program, 66,902 pigs were slaughtered. The first outbreak in Rio de Janeiro was noted on May 13, 1978. Animals began to die on April 30, 1978, and after 13 days, 200 out of a herd of 1000 pigs had perished. Secondary outbreaks were detected in herds fed on slops. The southern part of Brazil was selected for the first stage of an eradication program. Recently, the Brazilian authorities announced the complete eradication of ASF from their country.

ASFV appeared in the Dominican Republic in February 1978, and the last outbreak was reported in July 1979. Positive serological diagnosis was made in August 1980. An eradication program (with compensation to the farmers) was instituted, and it bore excellent results (11), as the disease did not appear in the new pig populations.

The presence of ASF in Cuba was recognized in January 1980, with 56 outbreaks occurring in three provinces in the most easterly region. Once the disease was confirmed, a state of emergency was declared in the affected provinces, and a state of alert in the rest of the country. Slaughter of pigs in the affected areas and systematic disinfection of the farms was carried out. No virus-positive pigs were observed after March 1980 (8).

The most recent spread of ASFV was noted in Belgium (1985) and in The Netherlands (1986) (C. Terpstra, personal communication). In these outbreaks, the first pigs infected were those fed with uncooked waste collected in ports and airports. These recent incidences were contained after slaughter of the infected animals and their contacts. In The Netherlands, ASF was diagnosed very promptly; the affected region was placed under strict control, and movement of pigs within the region was prohibited. This led to a successful eradication program.

The reports on the spread of ASF during the last eight years show that the virus was harbored in waste obtained from ports and airports and used as hog feed. A delay in diagnosis allows ASFV to spread from farm to farm. A further delay in the initiation of an eradication program results in a higher number of pigs having to be destroyed, with a concomitant increase in the cost of the operation.

Since ASFV seems to be resistant to climatic conditions and remains active in food waste and garbage, thus being able to infect the highly sensitive domestic pig, the practice of feeding untreated waste to pigs must be prohibited. A reliable diagnostic procedure with the necessary reagents should be made available to all countries for application at the very start of an outbreak suspected to be ASF on the basis of the clinical features.

REFERENCES

1. Montgomery, R.E. On the form of swine fever occurring in British East Africa (Kenya colony). J. Comp. Pathol. 34:159-191; 243-262, 1921.

2. Plowright, W. Vector transmission of African swine fever virus in hog cholera/classical swine fever and African swine fever. EUR 5904 EN, Commission of the European Communities, pp. 575-587, 1977.

3. Sanchez-Botiza, D. Estudios sabre la peste porcina africana en Espana. Bull. Off. Int. Epiz. 58:707-727, 1962.

4. Manso-Ribeiro, J. and Rosa Azevedo, J.A. Réappartition de la peste porcine africaine (PPA) au Portugal. Bull. Off. Int. Epiz. 55:88-106, 1961.

5. Carnero, R., Gayot, G., Costes, C., Delclos, G. and Plateau, F. Peste porcine africaine; données epidemiologiques symptomalogiques et anatomopatologiques collectées en France en 1964 et pouvant servir de base au diagnostic clinique. Bull. Soc. Sci. Vet. Med. 76:349-358, 1974.

6. Balbo, S.M. and Iannizzotto, G. Epizootological investigation, prophylaxis and eradication of African swine fever in Malta. African Swine Fever, EUR 8466 EN, Commission of the European Communities, pp. 42-46, 1983.

7. Contini, A., Cossu, P., Rutili, D. and Firinu, A. African swine fever in Sardinia. Ibid, pp. 1-6.

8. Negrin, R.E.S. Development in Cuba of a program for eradication of African swine fever in 1980. Ibid, pp. 36-41.

9. de Paula Lyra, T.M. Programme of control against swine fever in Brazil. Ibid, pp. 25-35.

10. de Paula Lyra, T.M., Pavez, M.M. and de Morais Andrade, C. Serological study of African swine fever in pig populations of Southern Brazil. Ibid, pp. 47-62.

11. Rivera, E.M. African swine fever in the Dominican Republic. Ibid, pp. 17-24.

12. Ordas, A., Sanchez-Botiza, C., Bruyel, V. and Olias, J. African swine fever: The current situation in Spain. Ibid, pp. 7-11.

13. Botiza, C. Reservorics del virus de la peste porcina africana. Bull. Off. Int. Epiz. 60:895-899, 1963.

14. Vigario, J.D., Castro Portugal, F.L., Festas, M.B. and Vasco, S.B. The present state of African swine fever in Portugal. African Swine Fever, EUR 8466 EN Commission of the European Communities, pp. 12-16.

Index

156